Structural Realism

THE WESTERN ONTARIO SERIES IN PHILOSOPHY OF SCIENCE

A SERIES OF BOOKS IN PHILOSOPHY OF MATHEMATICS AND NATURAL SCIENCE, HISTORY OF SCIENCE, HISTORY OF PHILOSOPHY OF SCIENCE, EPISTEMOLOGY, PHILOSOPHY OF COGNITIVE SCIENCE, GAME AND DECISION THEORY

VOLUME 77

Elaine M. Landry · Dean P. Rickles
Editors

Structural Realism

Structure, Object, and Causality

Springer

Editors
Elaine M. Landry
University of California
Davis California
USA
emlandry@ucdavis.edu

Dr. Dean P. Rickles
University of Sydney
Unit for History & Philosophy of Science
Sydney New South Wales
F07 Carslaw Bldg.
Australia
dean.rickles@sydney.edu.au

ISSN 1566-659X
ISBN 978-94-007-2578-2 e-ISBN 978-94-007-2579-9
DOI 10.1007/978-94-007-2579-9
Springer Dordrecht Heidelberg London New York

Library of Congress Control Number: 2011943742

Printed on acid-free paper

Springer is part of Springer Science+Business Media (www.springer.com)

Preface

It still seems reasonable to assume that scientific realism, in some form or other, is the majority position amongst both practicing scientists and philosophers of science. However, as Jarrett Leplin quipped some years ago, realism is "[l]ike the Equal Rights Movement . . . a majority position whose advocates are so divided as to appear a minority" (*Scientific Realism*, University of California Press, 1985, p. 1). Within the broader church of scientific realism, structural realism too has splintered into many diverse denominations, often underpinned by quite distinct motivations and argued for using quite different defences. There is, then, no monolithic position known as 'structural realism.' But there is a general convergence on the idea that a central role is to be played by relational aspects over object-based aspects of ontology—whether this reflects a *fundamental* fact about the world (the ontological thesis)—as opposed to our *engagement* with it (the epistemological thesis)—introduces still further divisions. Still further divisions *within* these divisions are created in the task of establishing what is actually meant by each of these theses.

Structural realism (if not quite labeled such) has been around in some form or other since at least as long ago as Poincaré's famous comment about Nature hiding the true natures of objects, leaving us instead with their their projections in the form of structural shadows. Perhaps if we looked closely enough, we could find earlier versions of it. Kant's theory of the division between *phenomena* and *noumena* certainly comes very close since phenomena will, by their very nature, involve relations (if only between observer and the qualitative properties of a system) as an inevitable consequence.

However, in more recent times, largely thanks to its integration with themes from the philosophy of physics (especially physical symmetries and group theory), it has evolved and broadened considerably from this earlier epistemological stance. For example, the position spawned an ontological version, according to which 'the shadows' are all that exists at a fundamental level. Whether objects can be entirely eliminated from one's ontology, in favour of a world of pure structure, is a subject of current debate, and one heavily discussed in several of the chapters that follow.

There is even, thanks to van Fraassen, an *anti-realist* (empiricist) counterpart to structural realism according to which one can agree with the structuralist aspects of structural realism but question the realist's characterisation of science in terms of *aim* thesis having to do with "grasping" (in something like the correspondence sense) the unobservable parts of the world, just as much as the observable parts.

There has also been some merging between the discussions of structure framed in the context of pure mathematics and those taking place in the physical sciences. Still more recently, structural realism has been applied within the social sciences. However, with this expansion of the forms and applications of structural realism there have emerged new challenges. It has reached the status of a mature position in the philosophy of science. But there are lingering issues stemming from the precise characterisation of structure, objects, and relations and the inclusion of causality and modality in structural realist positions.

How do we talk about causation in a purely structural world? Objects are usually viewed as the causes and the generators of change in the world; hence, given the structural realist's prioritization of relations over objects it is natural to ask about how causality fits into structuralism. Does the structural realist's world have sufficient resources to provide an adequate account of causal relations, or are objects a necessary condition of an adequate account of causation? What about deeper modal aspects of the world, as bundled into the concept of a law of Nature? Can such necessity supervene on an ontology of pure structure? This book is about these intertwined aspects, on the preliminary task of getting clearer on just what structural realists mean by *structure*, and how it relates to the objects that are part of our folk ontology.

The book is organized into three parts (though with much thematic overlapping amongst them):

- Part I examines the various frameworks (syntactic versus semantic) for, and the various interpretations (epistemic versus ontic versus methodological) of, structural realism, in addition to some of the standard motivations underlying these.
- Part II critically examines the differing frameworks, interpretations, and motivations of structural realism and examines their impact on what scientific theories say about the nature of objects and relations (and vice versa).
- Part III probes the issue of what can be said about what conception of object (if any) is needed to underwrite the concept of a causally connected, law-governed world.

This map corresponds roughly to the themed sessions of a 2007 workshop from which most of the chapters were drawn. However, in their conversion from talks to chapters, they have been significantly revised and updated and, we think, offer a faithful reflection of the current state of the art with respect to the problems and prospects of structural realism.

Acknowledgements

The contributions to this volume grew out of a workshop on structural realism held in Banff, Canada, August 28–30, in 2007. We are grateful to the Social Sciences and Humanities Research Council of Canada, the International Collaboration on Complex Interventions, and the University of Calgary for generous financial support for this workshop. Dean Rickles would like to thank the Australian Research Council for funding the Australian Research Fellowship under which this book was completed.

Contents

Contributors

Nora Berenstain University of Texas, Austin, TX, USA, nberenstain@mail.utexas.edu

Katherine Brading Department of Philosophy, University of Notre Dame, Notre Dame, 46556 IN, USA, kbrading@nd.edu

Anjan Chakravartty Department of Philosophy, University of Notre Dame, 301 Malloy Hall, Notre Dame, IN 46556-4619, USA, chakravartty.1@nd.edu

Steven French University of Leeds, Leeds, UK, s.r.d.french@leeds.ac.uk

James Ladyman Department of Philosophy, University of Bristol, Bristol, UK, james.ladyman@bristol.ac.uk

Elaine M. Landry Department of Philosophy, University of California, Davis, CA, USA, emlandry@ucdavis.edu

Antigone M. Nounou National Hellenic Research Foundation, Athens, Greece, amnounou@eie.gr

Stathis Psillos Department of Philosophy and History of Science, University of Athens, University Campus, 15771 Athens, Greece, psillos@phs.uoa.gr

Dean P. Rickles Unit for History & Philosophy of Science, University of Sydney, Sydney, NSW, Australia, dean.rickles@sydney.edu.au

Alexander Skiles Department of Philosophy, University of Notre Dame, Notre Dame, 46556 IN, USA, askiles@nd.edu

Ioannis Votsis Philosophisches Institut, University of Düsseldorf, Universitätsstraβe 1, 40225 Düsseldorf, Germany, votsis@phil.uni-duesseldorf.de

John Worrall Philosophy, Logic & Scientific Method, London School of Economics, London, UK, j.worrall@lse.ac.uk

About the Editors

Elaine M. Landry is Associate Professor of Philosophy at the University of California, Davis. emlandry@ucdavis.edu

Dean P. Rickles is Senior ARC Research Fellow at the University of Sydney. He is author of *Symmetry, Structure, and Spacetime* (Elsevier, 2007); editor of *The Ashgate Companion to Contemporary Philosophy of Physics* (Ashgate, 2008); and co-editor of *The Structural Foundations of Quantum Gravity* (Oxford: Clarendon Press, 2006) and *The Role of Gravitation in Physics: Report from the 1957 Chapel Hill Conference* (Max Planck Research Library for the History and Development of Science, 2011). dean.rickles@sydney.edu.au

About the Authors

Nora Berenstain is a PhD student in philosophy of science at the University of Texas at Austin. nberenstain@mail.utexas.edu

Katherine Brading is William J. and Dorothy K. O'Neill Collegiate Professor of Philosophy at the University of Notre Dame. She is co-editor of *Symmetries in Physics: Philosophical Reflections* (Cambridge University Press, 2003). kbrading@nd.edu

Anjan Chakravartty is Professor of Philosophy at the University of Notre Dame. He is author of *A Metaphysics for Scientific Realism: Knowing the Unobservable* (Cambridge University Press, 2007). chakravartty.1@nd.edu

Steven French is Professor of Philosophy of Science at the University of Leeds. He is author of *Science: Key Concepts in Philosophy* (Continuum, 2007); co-author of *Identity in Physics: A Historical, Philosophical, and Formal Analysis* (Oxford University Press, 2010) and *Science and Partial Truth: A Unitary Approach to Models and Scientific Reasoning* (Oxford University Press, 2003); and co-editor of *The Structural Foundations of Quantum Gravity* (Oxford: Clarendon Press, 2006) and *Correspondence, Invariance and Heuristics: Essays in Honour of Heinz Post* (Springer, 1993). s.r.d.french@leeds.ac.uk

James Ladyman is Professor of Philosophy at the University of Bristol. He is author of *Understanding Philosophy of Science* (Routledge, 2002), and co-editor of *Every Thing Must Go: Metaphysics Naturalised* (Oxford University Press, 2007). james.ladyman@bristol.ac.uk

Antigone M. Nounou is a researcher for the National Hellenic Research Foundation. amnounou@eie.gr

Stathis Psillos is Professor of Philosophy of Science and Metaphysics at the dept of Philosophy and History of Science in the University of Athens, Greece. He is the author of: *Knowing the Structure of Nature* (Palgrave 2009), *Philosophy of Science A-Z* (Edinburgh University Press, 2007); *Causation and Explanation* (McGill-Queens U.P. 2002); and *Scientific Realism: How Science Tracks Truth* (Routledge, 1999). He is also the co-editor of *The Routledge Companion to Philosophy of Science* (Routledge 2008). psillos@phs.uoa.gr

Alexander Skiles is a PhD student in Philosophy at at the University of Notre Dame. askiles@nd.edu

Ioannis Votsis is a researcher in the Philosophisches Institut at the University of Düsseldorf. votsis@phil.uni-duesseldorf.de

John Worrall is Professor of Philosophy of Science at the London School of Economics. He is editor of *The Ontology of Science* (Dartmouth, 1994); co-editor of *Rationality and Reality: Conversations with Alan Musgrave* (Kluwer Academic Publishers, 2006) and *I. Lakatos: Proofs and Refutations* (Cambridge University Press, 1976); and managing editor of *I. Lakatos: The Methodology of Scientific Research Programmes: Philosophical Papers Volume I* and *Mathematics, Science and Epistemology: Philosophical Papers Volume II* (Cambridge University Press, 1978). j.worrall@lse.ac.uk

Part I
Frameworks for Structural Realism

Chapter 1
The Presentation of Objects and the Representation of Structure

Steven French

1.1 Introduction: Presentation vs. Representation

The rise and development of various forms of structuralism (including structural realism and structural empiricism) has seen a concomitant array of diverse representations of structure placed on the table, from Ramsey sentences and mathematical equations, to the group-, set- and category-theoretic [35, 36]. This diversity has contributed to the confusion over what is intended by 'structure', particularly in the debate between realists and anti-realists and my intention in this paper is to help clarify the situation by drawing on Brading and Landry's distinction between the *presentation* of putative objects via the relevant 'shared structure' that our theories make available and the *representation* of such objects (as features of the world) by those theories. The obvious question then is: how is this (shared) structure represented? In addressing it I shall draw on earlier work that both examined the role of group-theoretic structure in the development of quantum physics and represented that role within the set-theoretic framework of the partial structures approach. Brading and Landry argue that the latter is surplus to requirements and that the relevant episodes can be understood from a 'minimalist' standpoint, structurally speaking [3]. However, I believe this rests on a misconstrual of the kinds of activity we are engaged in here, as philosophers of science, and a failure to keep distinct the work that is done by group structure at the 'object' level of scientific practice, and the work that needs to be done at the meta-level of the philosophy of science by the set-theoretic approach, for example. In both cases group theory and set-theory, respectively, are used as representational devices by physicists and philosophers, also respectively, but from the perspective of the meta-level group theory also functions as the mode by which the relevant objects are presented to us. As far as the structuralist is concerned this presentation then affords the means by which these objects can be metaphysically reconceptualised in structural terms.

S. French (✉)
University of Leeds, Leeds, UK
e-mail: s.r.d.french@leeds.ac.uk

E.M. Landry, D.P. Rickles (eds.), *Structural Realism*, The Western Ontario Series in Philosophy of Science 77, DOI 10.1007/978-94-007-2579-9_1,

I shall also respond to two recent criticisms of the group-theoretic presentation of objects: the first argues that the so-called automorphism towers that can be generated within group theory presents the structuralist with a form of underdetermination as to which structure to choose; the second raises concerns about the connection between this group-theoretic presentation and objectivity. Finally I shall return to the issue of representation in order to offer a reconciliatory way forward.

1.2 Beasts of the Field

Let me begin with what I call 'the beasts of the field', by which I mean the kinds of features of scientific practice, broadly conceived, that a structuralist approach, also broadly conceived, is, or should be, attempting to capture. Among the more notable examples we can identify the following: the structure of theories; the theory-data-phenomena relationship; inter-theory relationships; mathematics-theory relationships; the (broadly) philosophical implications of modern science, particularly physics but also biology, for example. This is certainly quite a broad range of features and one might wonder whether any one framework can in fact capture all of them. Although a pluralist approach is conceivable, many philosophers of science would agree that some form of unitary framework, stance or perspective is preferable and the two most well-known and discussed examples are the syntactic, favoured by adherents of the so-called 'Received View' and the set-theoretic, defended by proponents of the 'semantic' or 'model-theoretic' approach. My own inclination is towards the latter and within that approach one can distinguish the more from the less formal. Examples of the former can be found in the work of Sneed, StegMuller, Moulines and others, who attempt to capture a wide range of features of practice in considerable and, in parts, quite complex, detail. Giere's work, on the other hand, represents an example of the latter, where to a considerable degree the explication of theory structure etc. is left implicit and the focus is on the details of the relevant practice. Although I can appreciate the virtues of both ends of the spectrum, as it were, I prefer to occupy the middle ground, perhaps beside van Fraassen, where technical results can be deployed for the purposes of achieving our aims within the philosophy of science (this is a central point in what follows) but not to the detriment of understanding the relevant feature of practice we're trying to capture.

In a series of papers, my co-authors and I have tried to articulate the manner in which the partial structures approach can offer some form of middle-way between the above extremes. Thus we claim that it provides an appropriate representation of both theories and models, particularly with regard to their open-ended nature and the manner in which they can be further developed [12]; that, appropriately extended to include partial isomorphisms holding both 'horizontally' as it were, and 'vertically', it can capture the relationships between theories and between them and data models [4, 5, 12] and that further extended again to include partial homomorphisms, it can also capture the relationship between such theories and the mathematics in which they are 'framed' [7]; in particular and with regard to that last point, this approach can capture what Redhead famously called the 'surplus structure' of mathematics, which has played an important heuristic role in scientific developments [49].

Furthermore, in his now classic paper on the 'ontic' form of structural realism, Ladyman identified the partial structures framework as the appropriate mode of representation for this form, since it wears the relevant structural commitments on its sleeve, as it were [34, 36]. In particular, and in addition to the usual arguments that can be given in favour of the semantic approach, the significance for the ontic position of responding to theory change through the history of science provides further support for adopting an approach, such as the above, with its associated partial isomorphisms, as a way of capturing the relevant features of such change. Having said that, it has been noted that it may be less appropriate as a means of capturing the implications of modern physics, particularly where these involve the metaphysics of objecthood, given its obvious commitment to set-theoretic elements. I shall review ways in which one can respond to this concern below.

I have emphasised the role of this approach as a 'mode of representation' and it is important to reiterate that, on my view, adopting it does not entail that either theories or the structures they put forward as 'out there' in the world should be regarded as inherently set-theoretic in any way. A considerable amount of criticism of the semantic approach in general has been generated by taking it to imply that theories *are* set-theoretic structures, just as the syntactic view is supposed to regard them as logico-linguistic in some sense. As da Costa and I tried to make clear [12, p. 26] this is not our view; rather we take an important leaf out of Suppes' book in adopting a dual perspective: with regard to the 'extrinsic' perspective, theories should be characterised in set-theoretic terms which capture the relevant inter-relationships, as indicated above; from the point of view of the intrinsic perspective, they can be characterised in terms of sets of propositions, in terms of which one can talk of beliefs and—all importantly for the realist of course—truth [21]. That still leaves the question of what theories *are*, *qua* objects themselves, and although I have previously adopted a quietist line that suggests that this is not an appropriate question for the philosophy of science [22], more recently I have begun to explore a metaphysically nihilist approach which argues that there are no such things as theories, *qua* (of course) metaphysical objects [29]. The important point, however, is that as far as I am concerned, set theory offers an appropriate representational device for the philosopher of science and for the structural realist in particular; its use should not be taken to imply a particular ontological stance with regard to either theories themselves or the nature of the structure 'in' the world the realist takes them to represent.

1.3 Methodological Minimal Scientific Structuralism

Brading and Landry, however, have criticised the above approach as (meta-) methodologically unnecessary [3]. In its place they offer a form of 'methodological minimal scientific structuralism' that rejects these kinds of unitary frameworks at the meta-level, arguing that all that we need is an appropriate grasp of the relevant 'shared structure' at the object level of scientific practice, where theories themselves sit. In particular, with regard to the ontological claims of structural realism and indeed, of realism in general, they state:

> What we call minimal structuralism is committed only to the claim that the kinds of objects that a theory talks about are presented through the shared structure of its theoretical models and that the theory applies to the phenomena just in case the theoretical models and the data models share the same kind of structure. No ontological commitment—nothing about the nature, individuality, or modality of particular objects—is entailed. [3, p. 577]

Furthermore, they insist that,

> ... neither the framework of the semantic view of theories nor the appeal to shared structure alone offers the scientific structuralist a quick route to representation. [3, p. 580]

On this point we can certainly agree, since both representation in particular and structuralism in general may include further elements that in turn may be regarded as non-structural in certain senses. The concern then is whether the incorporation of such elements can be taken to undermine the structuralist programme and elsewhere I have argued that in relevant cases they do not [20]. Thus, it is clear that certain constraints must be imposed within the structuralist framework, without which it is not meaningful to talk of representation in the first place [28]. In one sense these constraints do represent significant non-structural elements, insofar as they embody theoretical content going beyond the pure logico-mathematical structure, which is linguistically specified and thereby constrains the possible systems in the world that are taken to be represented. However, the structure that the structural realist is concerned with should not be, and should never have been, construed as 'pure' logico-mathematical structure [20]; it was always intended to be understood as *theoretically informed* structure. Although the linguistic specification of these constraints may suggest that the structuralist account of representation is not purely structural, this theoretical content was always regarded as an inherent feature of the ontic form of structural realism to begin with (see French and Ladyman's reply to Cao in their [26]).

Returning to minimal structuralism, a crucial question is how do we make precise this concept of 'shared structure'? According to the partial structures approach, the answer is straightforward: 'Shared structure' is, or is represented by, (partial) set-theoretical structures plus the associated (partial) iso/homo-morphism. Landry, however, offers a more general view according to which shared structure need not be shared set-structure: the shared structure can be made appropriately precise via the notion of a morphism and the context of scientific practice determines what kind of morphism [38].

Thus she insists that,

> ... mathematically speaking, there is no reason for our continuing to assume that structures and/or morphisms are 'made-up' of sets. Thus, to account for the fact that two models share structure we do not have to specify what models, qua types of set-structures, are. It is enough to say that, in the context under consideration, there is a morphism between the two systems, qua mathematical or physical models, that makes precise the claim that they share the appropriate kind of structure. [38, p. 2]

Furthermore, she writes,

> I want to distinguish between semantic accounts that consider what the concept of shared structure is (what the appropriate type of structure is for formally framing the concept of shared structure in terms of some type of morphism) and those that consider what the presence of shared structure tells us (what the appropriate kind of structure is for characterizing the use of shared structure in terms of some kind of morphism as determined by some context), and to place focus on the latter. [38, p. 8]

1.4 Return to Group Theory

The case study Landry considers is that of the role of group theory in quantum mechanics. This is interesting in a number of respects, not least because of what it reveals about the relationship between physics and mathematics and the way in which the latter came to shape, in fundamental ways, the former. From the structuralist perspective it is particularly significant as both Cassirer and Eddington drew on group theory to underpin their respective structuralist positions [10, 18]. In both cases these positions have tended to be overlooked in favour of Russell's in particular (drawn upon by both advocates and critics of structural realism), even though at the time Russell had only a weak grasp of the newly emerging quantum mechanics and, it would seem (and as Eddington alleges) no appreciation of the significance of group theory for explicating an appropriate notion of structure. There are also interesting historical issues to consider with regard to the transition from the essentially geometric context of group theory as it applied to General Relativity (and as famously emphasised by Cassirer in particular) to its application to quantum theory. However it is the latter that I shall be concerned with here.

We recall [6, 16, 17] that one can broadly identify two intertwined strands in this episode ([41]; see also [1]): the 'Weyl programme' was concerned with the group-theoretic elucidation of the foundations of quantum mechanics, whereas the 'Wigner programme' focussed on the utilisation of group theory in the application of quantum mechanics itself to physical phenomena (both Weyl and Wigner contributed to both programmes). A crucial stimulus for the introduction of group theory was quantum statistics and, in particular, the connection between such statistics and the symmetry characteristics of the relevant states of the particle assemblies, arising from the non-classical indistinguishability of the particles. The fundamental relationship underpinning this move is that between the irreducible representations of the group and the subspaces of the Hilbert space representing the states of the system, with the group 'inducing' a representation in system space [55, p. 185]. Thus under the action of the permutation group, in particular, the Hilbert space of the system decomposes into mutually orthogonal subspaces corresponding to the irreducible representations of this group. These include the symmetric and antisymmetric, corresponding to Bose-Einstein and Fermi-Dirac statistics respectively, as well as those corresponding to so-called 'parastatistics'.

As well as possessing permutation symmetry, an atom is also symmetric with regard to rotations about the nucleus (if inter-electronic interactions are ignored) and again group representations can be used to label the relevant eigenstates. Weyl's mathematical work on the complete reducibility of linear representations of semi-simple Lie algebras allowed the irreducible representations of the three-dimensional pure rotation (or orthogonal) group to be deduced as well as the so-called 'double valued representations' representing spin [57, pp. 157–170]. As I have noted previously, there are two important features of this case [16]: first of all behind these 'surface' relationships lie deeper, mathematical ones. Thus the reciprocity between the permutation and linear groups [55, p. 281] not only functioned as 'the guiding principle' in Weyl's work [55, p. 377] but acted as a 'bridge' within group theory. Practically it was also significant since continuous groups can be more easily handled than discrete ones. Hence the appropriate representational framework in which to situate the mathematics-science relationship in this case should incorporate families of structures on each side. The application of group theory to quantum physics crucially depends on the existence of this bridge between structures within the former.

Secondly, both group theory and quantum mechanics were in a state of flux and development at this time and the structures should be regarded as significantly open in certain respects. As has been repeatedly emphasized, the partial structures programme appropriately captures this feature at the appropriate representational level. From such a perspective, both mathematical and scientific change can be treated as on a par at the 'horizontal' level as it were and, 'vertically', given the *partial* importation of mathematical structures into the physical realm in this case, partial homomorphism provides the appropriate characterisation of this relation [7]. It is by precisely accommodating and, thereby, presenting such features that a representational framework such as that provided by partial structures proves its worth.

The value of such an approach is further exemplified by Wigner's subsequent group-theoretic development of isospin based on an analogy between atomic and nuclear structure that is both partial and dependent on certain idealisations [17, 41, pp. 254–278]. Drawing on Heisenberg's treatment of the forces between protons and neutrons by analogy with his earlier account of the exchange forces in the ionised hydrogen molecule, Wigner [56, p. 106] took both these forces and the masses of the particles to be approximately equal which allowed him to treat them as indistinguishable (apart from their charge). They could then be conceptualized as two states of a new kind of particle, the 'nucleon'. The kinds of idealisations can be represented via partial isomorphisms holding between the partial structures [17]: taking them in stages we move from protons and neutrons with nonequal forces, to a model with protons and neutrons and equal forces, to one of nucleons. Merging them together, the fundamental idealisation is the shift from protons and neutrons to the nucleon and in this way the nucleus can be treated as an assembly of indistinguishable particles. By analogy with the situation in the atom this in turn suggests the introduction of a further symmetry group on the back of the analogy between representations of nucleons and representations of electron spin: the relevant decomposition of the Hilbert space is analogous to the decomposition of the corresponding Hilbert space for the spin of an electron (the relevant groups have isomorphic Lie algebras).

Within the representational framework, we have an isomorphism between the anti-symmetrized tensor power of the direct sum of two Hilbert spaces and a direct sum of products of anti-symmetrized tensor powers which reduces the problem of determining the interaction between the protons and neutrons to that of considering 'particles' of the same kind, the Hilbert space of each of which is the direct sum of the proton and neutron Hilbert spaces [41, pp. 257–258]. The analogy between atomic and nuclear structure thus reduces to that which holds between the relevant anti-symmetrized Hilbert spaces for a system of electrons in an atom and a system of nucleons in a nucleus. However, the analogy is multiply incomplete (ibid., p. 259): the proton/neutron decomposition does not depend on choosing an 'axis'; both protons and neutrons also have spin 1/2 [56, p. 107] and so the representations of the rotation group in the relevant Hilbert spaces are irreducible in the electron case but the direct sum of two equivalent irreducible representations in that of the nucleons. Thus the introduction of isospin, on the physics side, requires, on the mathematical side, the use of an appropriate symmetry group that is more complicated than in the atomic case since the corresponding Hilbert space is of higher dimension [41, p. 259]. This prompted Wigner to move to the representations of the four-dimensional unitary group U(4), which yields, instead of multiplets, the 'supermultiplets' of nuclei [56, pp. 112–113].

The partial structures framework nicely captures this incomplete analogy between atomic and nuclear structure. Following Hesse's classic division, there is a positive analogy that holds between the atom with its electrons and central nucleus and the nucleus itself, with its nucleons and centre of gravity. There is a further two-fold analogy between the treatment of the nuclear particles as indistinguishable and the indistinguishability of the electrons; and also between the spin of the electrons and the isospin of the nucleons. The application of the permutation group then follows on the back of the former. With regard to the latter, the positive analogy holds between the direct sum decompositions into the relevant sub-spaces. The negative analogy is likewise two-fold: there is no 'axis' of isotopic spin in the nucleon case but more profoundly, the relevant Hilbert space is of a higher dimension since both protons and neutrons also have spin. Thus the deeper disanalogy between the two structures concerns the replacement of the rotation group by the four dimensional unitary group U(4). Isospin then went on to become an important feature of elementary particle physics, as the relevant structures were extended via the neutral analogy. As is well known, it was through efforts to combine the SU(2) group of isospin and the U(1) group of strangeness or hypercharge that SU(3) was proposed as the group of the quark model. Isospin then ceased to be regarded as 'fundamental', and with the development of colour and the electroweak group, so did SU(3) [40].

The point, then, of the above summary is to illustrate the advantages of adopting an appropriate representational framework such as that offered by partial structures. In particular, it allows us to re-describe and re-present the relevant historical elements in terms that are accessible to the philosopher, such as 'positive analogy', 'partial isomorphism' and so on. Furthermore, although this re-presentation will display the work performed by group theory itself, it is clear that this features at the 'object' level, as it were, of scientific practice, not at the level of the representations of philosophers of science; I shall return to this point below.

However, the crucial claim that Landry draws from the application of group theory to quantum mechanics is that,

> ... what does the real work is not the framework of set theory (or even category theory); it is the group-theoretic morphisms alone that serve to tells us what the appropriate kind of structure is. [38, p. 11]

More generally, she claims, it is the use of the concept of shared structure that determines the kind of structure and characterises the relevant meaning and all the relevant work is done by the contextually defined morphisms (as we shall see, what counts as the relevant 'work' in these cases is crucial).

1.5 Implications for Structural Realism

Before responding to the above claims, let us consider a fundamental question for structural realists: which structures in the physical world are we to be realists about? There is a tradition, as I have already indicated, going back to Cassirer, Eddington and others that takes these structures to be represented by group-theory (of course, Cassirer and Eddington were not realists about these structures, at least not in the way we currently understand the term).

The case of spin is illustrative in this regard. Morrison provides a useful summary of the history, nicely mapping the intertwining of theoretical and experimental aspects [44]. Her conclusion is that spin is a 'hybrid' notion possessing both mathematical and physical features and 'bridges' (that word again!) the mathematical and physical domains. And this is revealed by the fact that it essentially drops out of the mathematical formalism (of the Dirac equation, underpinned by group theory), in the sense that it is required to secure conservation of angular momentum [44, pp. 546–547]. Morrison takes this hybrid character to pose a challenge for realism since the latter stance requires that an appropriate physical interpretation of this property be given and the manner in which the mathematical and physical are intertwined renders such an interpretation 'otiose' [44, p. 548]. Now, this is a strong claim that, if accepted, would push us either to drop standard realism or move toward some form of Platonism (again see [26]).

However, the idea cannot be that the simple combination of mathematical and physical features in the description of spin renders any interpretation otiose, since that is obviously true of many such properties in physics, nor that it is required in order to save the conservation of a quantity; rather it must be that the mathematical features are such that no purely physical interpretation is possible. As Morrison notes, '[o]ur current understanding of spin seems to depend primarily on its group theoretical description' [44, p. 552] so it is obviously this group-theoretic description that is problematic. What this yields, as Eddington long ago pointed out, is not simply a pattern of entities, or even a pattern of relations, but rather a 'pattern of interrelatedness of relations' [14, p. 278]. What group theory gives us, then, is the appropriate algebra of operators representing rotations acting on rotations, for which the 'pattern of interrelatedness' is manifested in the associated multiplication table. Presumably it is this that is resistant to a straightforward realist interpretation.

As Morrison notes, an obvious response would be to elaborate such an interpretation in structuralist terms but after briefly sketching the respective virtues of epistemic and ontic structural realism she unfortunately misunderstands the nature of the latter and concludes that it cannot help in this case, since,

> [o]n this account the structures become no less mysterious than the physical entities they have reconceptualised. To say that the mathematics is a description of the structures but that they themselves are something else leaves us in the precarious position of affirming the existence of a 'something I know not what'; structures whose natures are described in a certain way. But this was exactly the problem that ontic SR was designed to solve. [44, p. 554]

Now the first misunderstanding here concerns the supposed mysterious nature of physical entities: for the ontic structural realist this is because what Brading calls their 'individuality profile' is underdetermined (Chapter 5); that is we cannot tell whether they are individuals or not. The 'mystery' is resolved and the metaphysical underdetermination dissipated by reconceptualising such entities in structural terms, rather than as objects. This 'mystery' is entirely different from that which Morrison now associates with the structures. Here it has to do with the difference between the mathematical and the physical and the claim that however we understand the former (platonistically perhaps) the latter will be 'something else'. But if this is a 'mystery', it is surely one that arises for *any* form of realism since it has to do with appropriately characterizing the physical. And again the mystery is resolved by, for example, alluding to the causal nature of 'the physical', something that I would maintain can be accommodated within a structuralist framework [19].

The second misunderstanding concerns the assertion that 'the structures "are what they are"' [26, 44] which Morrison takes to amount—'without any justification or insight'—to a 'mysterious ontological claim'. But the context here is that of a response to Cao's accusation of closet Platonism as part of which we insist that ultimately 'the structure' is just the world and any further questioning can only be met with the reply 'it is what it is'! It may remain mysterious as to how we are to understand structure metaphysically but that does not seem to be what Morrison is alluding to and anyway, some progress is being made to articulate that sense of what structure is [19, 23]. Finally, she notes that, as previously pointed out, OSR eschews reference (as does ESR) and hence, 'even if we accept that we should understand spin in terms of structure, [structural realism] doesn't help us to determine the reality of that structure.' [44] But of course, even if we inserted a reference relation into our position, it would not help us to 'determine the reality' of structure. And as it happens, attempts have been made to consider structuralist equivalents of reference [21, 28].

However, there is a more general disagreement that emerges in the following claim:

> Adding a layer of metaphysics answers none of the questions about the nature of spin that aren't already implicit, or indeed explicit, in the physico-mathematical description provided by quantum theory. To reconceptualize that description in terms of a metaphysics of *sui generis* structures renders the problem more convoluted. Nor do the activities associated with experimental detection become more perspicuous when understood in terms of these unexplained structures. [44]

Here the issue arises of how much metaphysics the realist in general and structural realist in particular should allow into her position [24], but whatever Morrison's views on that fundamental matter, it seems odd to raise a crucial problem for realism, then when attempts are made to solve that problem through the deployment of metaphysically interpreted structure, to insist that all the important features of the nature of spin are already implicit in the very physico-mathematical description that generated the problem! Again, the task of the structural realist is not to reconceptualise in terms of a metaphysics of *sui generis* structures, but rather to do so in terms of an account of such structures appropriately metaphysical understood. And there is an analogy here between Morrison's insistence on remaining with the object-level's physico-mathematical description and Brading and Landry's: the response in both cases is to insist right back on the significance of appropriate representational devices that give content to our realism and philosophy of science respectively.

As for the activities associated with experimental detection, there are two things the structural realist can say. The first is that one might hope that shifting away from a metaphysics of (individual) objects and their associated (typically monadic) properties would in fact help introduce further philosophical perspicuity into these activities. The second is that the kinds of experimental traces we usually observe (tracks in a cloud chamber etc.) are typically taken to support the exportation into the micro-realm of an inappropriate metaphysics, namely that of macro-objects and their supposed properties. It may be that a fully structuralist understanding of both science's theoretical and experimental activities is precisely the way to go in order to achieve an overall and fully harmonious metaphysics of science.

Moving on, I completely agree with Morrison when she states that '[p]art of the difficulty with attempts to generate a physical notion of spin concerns the way the electron is pictured in the hydrogen atom as a quantum mechanical object.' [44] We are led astray by this fundamentally object-based metaphysics to view spin as rotation around an axis, as described by relations between observables. But these relations are represented by operators and as Eddington perceived it is the algebra of these operators that describes the structural 'pattern of interrelatedness of relations'—unpacking the latter will then give us our metaphysical interpretation. This goes beyond simply acknowledging that, following Wigner's account of elementary particles, spin is simply a group invariant characterising the unitary representation of the Poincaré group associated with the wave equation. It is the group-multiplication table that represents the structure in this case and the metaphysics of the latter will be shaped by the features of this table.

This tradition of focussing on group-theoretic structure has been identified with a variant of structural realism, called 'Group Structural Realism' ([50]; see also [32]), although the significance of group structure is so intimately bound up with ontic structural realism in the works of myself and Ladyman that one may wonder whether the view really deserves a separate designation. Nevertheless it is precisely the tension between this group-theoretic informing of our understanding of structure and the set theoretic framework of OSR that Landry exploits in her critique. Before returning to tackle this, let me address two recent objections to this emphasis on group structure within the structural realist stance.

1.6 Objection 1: Toppling the Tower of Automorphism

In articulating his version of 'Group Structural Realism', Roberts identifies what he calls the 'higher structures problem' [50]. The discussion kicks off by describing French and Ladyman as raising concerns over whether structures might be 'describable' in terms of structures [19, 27]. In fact what we were concerned with was responding to Psillos's worry that if we incorporate causality within our structuralism, and if causality is itself understood in structural terms, then we obtain a kind of 'hyperstructuralism' leading to an infinite regress. The issue is whether the causal empowerment that some might claim is inherent in the structure of the world is itself describable in structural terms. The dilemma is that if it is, a regress threatens; but if it isn't, then a non-structural element has been introduced into our ontology and more generally our structuralism is limited. Note that this is a metaphysical issue to do with Psillos's fundamental worry whether causality can be understood in structural terms. Our response is that neither horn is particularly problematic. On the one hand, one can swallow the infinite regress. That would have obvious implications for any presumption of metaphysical fundamentality but Saunders has already suggested the possibility of a non-grounded form of structural realism in his slogan 'It's structures all the way down' [51] and the lack of such fundamentality can certainly be accommodated [40]. On the other, one can accept causality as metaphysically non-structural, but then its not clear that that undermines the structuralist project in any fundamental way if one takes the core of that project as the reconceptualisation or elimination of *objects*. One can argue that structural realism is not fatally weakened by having removed objects from its metaphysics and then admitting causal powers as non-structural features. The issue here again has to do with how 'pure' we are expected to take our structuralism to be [20].

But this has to do with the structural describability of *causality*, whereas what Roberts is concerned with is the structural describability of structure. This of course is unproblematic in that one can appeal to structure to describe and represent structure; indeed, given, for example, Eddington's emphasis on the relevant structure as understood in terms of the interweaving of and hence relations between relations, such a representational move lies at the core of the form of structural realism defended here. However, Roberts sees this describability function as generating a dilemma.

Thus he articulates Group Structural Realism in hierarchical terms, with group theoretic structure sitting at the topmost or most fundamental level. He then asks the question, 'If S is a structure, what is the status of the structure of S itself?' Now some care needs to be taken, as a lot could hang on what is meant by status here. Roberts takes it to mean 'metaphysical' status, rather than representational, say and uses this to generate the dilemma:

> On the one horn, we would like to choose just one structure to be at the top of our metaphysical hierarchy. But it is unlikely that we will be able to give a well-motivated reason to choose between a structure S , and the structure of S itself. This pushes us to the other horn: we must promote the whole shebang, both S and the structure of S, to a metaphysically 'fundamental' status. But this account of metaphysics, if one can even make sense of what counts as the 'whole shebang,' leads to an much more complex hierarchy, which need not satisfy the aims of structural realism. [50, p. 57]

The concrete example he chooses is the automorphism group $Aut S$ of S and the dilemma bites like so: either we give a reason for choosing $Aut S$ over S as more fundamental (or vice versa) or we swallow the 'whole shebang', but that's a big shebang, given the existence of so-called automorphism towers; that is, a succession of automorphism groups of automorphism groups that are non-trivial in the sense of generating new groups and that may only terminate in the transfinite, or even cycle.

Let us begin with horn number 1: one approach might be to adopt a variant of Redhead's point that whenever we take the physical structures we're interested in and embed them into 'higher' mathematical structures, we obtain a lot of surplus structure that may or may not be heuristically very useful (for a discussion of surplus structure in the context of structural realism see [25]. This embedding can be represented set-theoretically (via the notion of partial homomorphism; [7]) and although Roberts is obviously correct that the situation with the automorphism towers cannot be represented in this way (since AutS is not a sub-structure of S), nevertheless the issue is the same: where do we draw the metaphysical line between those structures we take to represent the world and those that are surplus?

One option would be to appeal to mathematical considerations but as Roberts points out, the tower can be extended downwards and in different ways, and a kind of underdetermination arises. Again, this seems little more than a reiteration of the point that mathematics yields surplus structure, which hardly comes as a surprise and the issue remains as to how to draw the relevant line.

Thus an alternative but obvious option is to draw that line on physical grounds, by appealing to the objects to be represented. However, Roberts insists this leads to circularity, because, he states, the group 'provides' physical objects with their properties, so we can't appeal to those objects to pick out the group. This seems a confusing way of putting things. What 'picks out' the group is the relevant theoretical context via the usual justificatory moves (and thus grounded in the appropriate empirical context). The structural realist then metaphysically reconstitutes any putative physical objects in group-theoretic terms, claiming that is the latter that articulates the sense of structural reconceptualisation or elimination of these objects. Thus rather than a circle, we have two 'arms': one justificatory and hence epistemological; and the other metaphysical and hence ontological. (As we shall see, a failure to note the justificatory side of things undermines the further criticisms from Debs and Redhead to be discussed below).

A related option is to stick with whatever is closest to the physics, by adopting what Roberts calls the 'natural physical attitude'; that is we accept the group that is most naturally suggested by the physics [50, p. 65]. That in effect is what Brading and Landry would have us do and leave it at that. But as Roberts indicates, the structural realist insists this is not enough. Leaving aside the issue of the representation of the relevant results at the meta-level of the philosophy of science, there is the question of how the physics is to be interpreted. By interpretation here I mean 'situating' the mathematically informed physics of $SO(3)$, say, in the context of an appropriate metaphysical understanding of structure. Roberts reminds us that we should not read the metaphysics off the physics but that of course is not what I am suggesting here. We should not confuse interpretation with 'reading off' and given that the whole point of ontic structural realism is to come up with a metaphysics that 'best' fits the physics, in the sense of being metaphysically minimalist and avoiding object-talk it is hard to see how the structuralist could be 'barred' from appealing to interpretation here. Again, the move is to take what the physics gives us, as it were, as revealing what the structure of the world is like (so for example, the claim might be that that structure can be represented, at the object level by $SO(3)$) and interpreting that via an appropriate *metaphysical* understanding of structure (here Roberts' claim to have provided such an understanding in clear terms indicates that he has fallen under the same spell of object-level 'shared structure' as Brading and Landry; although its group-theoretic presentation may help inform our metaphysical understanding, it is not sufficient).

Roberts also considers seizing the second horn and simply accepting the whole 'tower', as it were. The obvious worry here is that this is just too 'wild' and ontologically extravagant. Now of course if one were to reject the kind of hierarchical framework that Roberts assumes, where there has to be a fundamental structure underpinning all the rest, then this worry might dissipate. It is important to note however, that what we have is not physical structures represented by different groups all the way down, as Saunders suggested, but a tower of mathematical excrescences associated with the one group (e.g., $SO(3)$). Are all of these mathematical objects to be seen as further features of the structure of the world? That does seem ontologically inflationary. But again one can see this as a consequence of the surplus structure that mathematics inevitably provides and we return to drawing the line in terms such as the above. The fundamental point is that we have to draw the line anyway since the structure we are realists about is physical not mathematical.

1.7 Objection 2: In Defence of Invariantism

The relationship between symmetries with their associated invariants and objectivity has long been noted. Weyl, for example, famously took objectivity to mean invariance with respect to the relevant group of automorphisms (for a more recent discussion see [8, 53]. Nozick [45] has strengthened this line by claiming that invariance *explains* three crucial features that render a fact objective, namely:

1. It is accessible from different perspectives.
2. There can be intersubjective agreement about it.
3. It holds independently of people's beliefs, desires, observations, measurements.

And we come to know which are the relevant invariants through what he calls the 'bootstrap process of scientific investigation' (ibid., p. 84). What he means by this is some combination of heuristic and justificatory moves that lead us to those theories that the realists take to be our 'best explanations'. Recently Debs and Redhead have raised a series of criticisms against this argument [13], although they articulate these in terms of the heuristic role of symmetries only, thus ignoring the justificatory aspect of the process.

Thus their first criticism has to do with sorting out what is significant: they point out that symmetries come in various shapes and forms and that it is difficult, if not impossible to know beforehand which will be heuristically fruitful or not. This seems an obvious point but it hardly impacts on the kind of objectivity claim articulated above. The core claim of the criticism is that no account has been given either for why some symmetries are physical, others mathematical, some dynamical, others accidental etc, or for why some are fruitful and others not. Indeed 'history suggests' that no such account will be forthcoming. The significance of certain symmetries thus seems to be a 'brute fact'.

However, if you're a realist, then this significance, understood appropriately broadly, is 'explained' by the way the world is. This would be the ultimate 'brute fact'! If by significance is meant something akin to heuristic fruitfulness, then retrospectively we give the same answer—gauge invariance has turned out to be so fruitful because that's the way the world is structured—and prospectively, we can only say 'that's why its called heuristics' since we can't know ahead of time which will work, and we can't give an algorithm for scientific discovery. As for 'explaining' the above differences again we will have to appeal to the sort of 'line drawing' cited in response to Roberts' concerns. And again, when it comes to distinguishing a physical symmetry, as represented by a mathematical group that is applied, from a non-physical one, as represented by a group that is not applied, we simply have to refer to the structure of the world. Ultimately we have to stop somewhere in our explanatory endeavour and if the question is why one group represents the world and not another, the realist's answer will be that is the way the world is. If this seems unsatisfactory, I think it seems so for reasons that have nothing to do with the role of invariants in establishing objectivity.

Debs and Redhead's second criticism has to do with choosing what they call 'The Definitive Group'. Here the core concern is that 'different aspects of the physical world have different symmetries' but identifying the 'universal physical symmetries' is difficult. As an example they contrast the hydrogen atom with relativistic space-time, where we have two models structured by very different symmetry groups [30], suggest that the former is better represented via groupoids and relate these to an extension of the concept of symmetry in terms of equivalence classes. In the latter case, a kind of fruitful heuristic leapfrogging occurred but not in the former. Now this might be expected given the very different physical systems concerned, and of course sometimes structures and symmetries are exportable from one domain to another very different one (consider for example the renormalisation group in the context of the development of quantum field theory). It might well be that a set of symmetries applicable to one system turns out to be applicable to another very different kind of system. As Debs and Redhead say, ultimately this is determined on a case-by-case basis and it is empirical success that plays a fundamental role in this determination, but of course, no one but the sociologists of science expected it to be determined in any other way.

Thus if one is a convergent or non-pluralist realist, one will insist that ultimately we will arrive at the set of fundamental 'universal physical symmetries'. Debs and Redhead claim that the problem is how to pin down this set, when all we have to go on is their heuristic fertility. But of course we don't just have that, we also have empirical success and although perhaps a complicated story will need to be told about how that flows up from the phenomena to the symmetry principles, that is surely not insurmountable. So the answer to their question 'If these symmetries are so selected due to their heuristic effectiveness, then why add to this the notion that they are associated with objectivity?', is that they are *not* so selected and the 'adding to' here simply reflects the difference between heuristics and justification.

Their third criticism is essentially that invariantism is tied to the search for a unified theory but objectivity should be something that is independent of such a goal. The connection is made explicit in their claim that the search for symmetry may be linked to the search for increasingly unified theories [13, p. 71] but that's a big 'may be'. One could presumably still be an 'invariantist' and a Dupré style pluralist or a Cartwrightian dappler—each domain or 'patch' would have its own set of symmetries in terms of which objectivity would be given. Certainly, one could still retain subject-independence within this framework and a Cartwrightian would surely object to the claim that nothing could be more subject independent than a Grand Unified Theory (GUT). Similarly if one were an ontological non-reductionist, one would insist that each 'level' could have its own symmetries—if such could be made sense of (e.g., at the biological level).

Taking the standard convergent realist line (where convergence is to the GUT), it is still not clear why a problem arises. The worry seems to be that we could not have 'full' objectivity until the GUT is known, and 'partial' objectivity is unacceptable. But I don't see why that should be. Why can't I adopt a broadly fallibilist stance (towards my beliefs) that allows me to accept that at least some of what I currently take to be objective may turn out not to be—so parity goes out the window, to be replaced in some sense by CPT—incorporating specific partiality, so I have good grounds for believing that at least part of the set of facts represented by current theory are objective?

Debs and Redhead insist that according to the invariantist approach classical physics must fail in its objectivity because of the relevant lack of invariance—to which the appropriate response is surely 'yes, yes it does!' Again, I do not understand why it is 'overly restrictive' to say the models of classical physics are not objective, as they claim. We can still say they're pragmatically useful, approximately accurate within the appropriate limits etc., and even that they are partially or pragmatically true [12]. They are not objective because the theory is strictly false, and that determination is of course made within the domain of justification, retrospectively as it were. Prospectively, we still want to retain the capacity to make objective claims that are provisional but there's nothing about invariantism or the supposed link to unification that prevents us from doing that. We just need to adopt the appropriate fallibilist stance(s). On such a stance, objectivity would indeed be an ideal, to be reached once we have the GUT, but I don't understand the criticism that a definition of objectivity must offer more than this to be useful—not least because the invariantist definition is offering more than this.

Debs and Redhead's aim is to dismiss (absolute) invariantism as a 'tantalizing illusion' and so set the stage for the entrance of their perspectival form, which includes elements of convention. I shall not pursue this further here but note that I am unmoved by their worry that the 'objective identities' of objects could be construed as objective features of some model and these are clearly not invariant. As an eliminativist I'm happy to see the objects and their identities disappear, so given my response to their arguments above, I remain a non-perspectival structural realist who takes objectivity to be appropriately grounded in the invariants that group theory presents.

Having dealt with these concerns arising from the placing of group theory at the heart of OSR, let us now return to Landry's claims.

1.8 Set-Theory as Cleaver

We recall her fundamental question: if it is group theoretical structures that we are going to be realists about, in the sense indicated above in the case of spin, then where is set-structure doing any real work? She insists that,

> ... if one wants ... to use this kind of structure as a tool to carve 'the world' into its 'natural kinds', then one cannot, in addition to claiming that group theory is 'the appropriate language', claim that all such group-theoretic kinds are set-theoretic types, *unless* one is ready to hold fast to, and provide justification for, the Bourbaki/Suppesian assumption that all scientifically useful kinds of mathematical structures *are* types of set-structures. Nor can one use this assumption to make a more robust, ontologically read, structural realist claim about the structure of 'the world', unless one wants to impose (or presume) that set theory cuts not only mathematics but indeed, Nature at its joints. [38, p. 15]

In other words, there is, first of all, a tension, at the very least, between the claim that group theoretic structure is what we should be realists about and the adoption of the set-theoretic approach by the ontic structural realist, and this tension can only be dissipated if we adopt the Bourbakian line. Furthermore, that latter response would propel us into the unsavoury position of claiming that the world is somehow set-theoretic, in an ontological sense.

Thus Landry urges that structural realism should free itself from its set-theoretic ties and adopt a minimalist form of structuralism based on this concept of shared structure, understood as that structure that is actually 'doing the work' in the relevant physical context. However, I suggest that this apparent tension is the result of confusion over the different representational roles being played by the respective structures and that, furthermore, there are advantages to retaining a set-theoretic representation of theories whilst also maintaining a group-theoretic presentation of structure. In particular I think we can easily resist falling into some form of set-theoretic Platonism about the world.

1.9 Presentation of Objects and Properties via Shared Structure

Let us consider briefly how physical objects are typically presented within theories. We might approach this informally via a journal or textbook presentation of the theory concerned, which might typically set out the fundamental principles, laws etc., together with some indication of that the theory is 'about'. Or we might adopt a more formal approach, either following the logical empiricists and reconstructing the theory in a formalized language, or, more moderately, offering an appropriate description in predicative terms [52]. Taking this route, the non-logical symbols of the relevant formal language are derived from the theory and interpreted in terms of physical properties, relations and functions. As Saunders puts it,

> ... we may read off the predicates of an interpretation from the mathematics of the theory, and because theories are born interpreted, we have a rough and ready idea of the objects they are predicates of. But there is nothing systematic to learn from the formalism to sharpen this idea of object. [52, pp. 290–291]

Saunders' concern here is with identity and indiscernibility in quantum physics and he draws on the 'purely logical aid' of a form of Leibniz's Principle of Identity of Indiscernibles given a Quinean twist by means of which quantum entities can be regarded as 'weakly discernible' and hence as objects in a 'thin' sense. This effectively hones the 'rough and ready' idea of an object in the quantum case into something more metaphysically robust (although still structural in a sense, since the weak discernibility is famously conferred via asymmetric relations). But as he notes, within the theory itself, identity signifies only the equality or identity of mathematical expressions, not of physical objects. Furthermore, the obvious worry the structuralist may have is that during the birth process, as it were, this rough and ready idea will be shaped by metaphysical preconceptions drawn from our interactions with 'everyday', macroscopic objects and inappropriately exported into the micro-realm described by modern physics.

Quine himself, of course, famously described physical objects as irreducible 'cultural posits' that are '...conceptually imported into the situation as convenient intermediaries not by definition in terms of experience, but simply as irreducible posits comparable, epistemologically, to the gods of Homer ...' [47, p. 44]. What the logical form of the relevant re-description gives us are the values of the variables that signify what exists, but ontological relativity implies that objects are nothing more than 'mere nodes' within the global structure that can be interpreted under widely different ontological frameworks while leaving the evidential base undisturbed. Since ontology is so plastic on this view, Quine concludes that structure is what matters, not the choice of objects (of course, although he refers to Ramsey and Russell he does not have structural realism in mind here). The very notion of an object, he insists, should be seen as a human contribution, resulting from our inherited apparatus for organizing the 'amorphous welter of neural input' (and hence one can draw connections with Cassirer's neo-Kantian view; [10]).

We can view objects as standardly presented in the context of the associated theories, either as part of this rough and ready understanding attached to the interpretation the theory is 'born' with (conceptually imported, as Quine puts it), or extracted from an (at least moderately) formal re-description of the theory, with the help of purely logical aids such as PII. What this then underpins is the standard metaphysical picture in which we 'build up' from the bottom, as it were, beginning with objects, which 'have' (in some sense) properties, that are then related in various ways, these relations captured and described by the laws associated with our theories. Thus as an example, a particle such as an electron, metaphysically regarded as an object, possesses the intrinsic property of charge, which 'enters' into relations with other instances of charge, these relations being then described by Coulomb's Law, say.

The structuralist offers a different 'top down' picture in which we start with the laws and principles 'presented' (on the surface as it were) by the theory, interpret these, at least minimally, in terms of relations and properties, but then resist the temptation to take that further metaphysical step and regard these last as possessed by (metaphysically robust) objects. On this view, these relations and properties are features of the fundamental structure and what we standardly designate as 'objects' are indeed mere nodes in this structure. In particular, elementary particles are not metaphysically robust objects under this perspective, but are reconceptualised structurally and represented by the relevant symmetry groups [8]. Historically, we can draw on Cassirer (stripping away the neo-Kantian elements of course), who argued that,

> ... that which knowledge calls its 'object' breaks down into a web of relations that are held together in themselves through the highest rules and principles. (Cassirer 1913, trans. in [31], p. 522)

These 'highest rules and principles' in turn are the symmetry principles that represent the invariants in the web of relations itself. It is these principles that are represented group-theoretically and hence the relevant group supplies the general conditions in terms of which something can be viewed as a putative 'object' ([10, 15, 32, 39] have also emphasised the role of symmetry in this context and in particular its ontological significance).

Of course there is much more to say about the structuralist view of laws for example [11], or the sense in which putative objects are dependent upon [23] or constituted by [2] laws, but the point I want to emphasise here is that within this 'top-down' picture putative 'objects' are presented (again, at the level of the philosophy of science), not as part of the birth pangs of the theory, nor as imported conceptual intermediaries, nor with the help of purely logical aids, but via the relevant symmetry groups. Brading and Landry take these to be captured by the relevant 'shared structure' and I certainly agree that this is context dependent in the sense that it is the physical context that 'reveals' that aspect of the world-structure. However, we need to be clear about what, or who, is doing the relevant work in these cases.

1.10 Doing Useful Work

So, recalling Landry's point that it is the use of the concept of shared structure that determines the kind of structure and that all the relevant *work* is done by the contextually defined morphisms, let's ask: who's using and what's working?

First of all, it is obviously the physicists/mathematicians who used and continue to use group theory in the relevant physical contexts, not (partial) set-structures (except maybe implicitly, if one were to insist that all mathematics is reducible to such structures!). In particular, in the context of the quantum revolution, it was group theory, not (partial) set-structures that was effectively doing the (physical, mathematical, and hence object-level representational) work. It was these group-theoretical structures that Weyl used to re-conceptualise the foundations of quantum mechanics and that Wigner used to obviate the need to consider specific forces and dynamical laws and that he deployed to underpin the notion of isospin. And as indicated, it is terms of these structures that we can consider putative 'objects' (taken, from the structural perspective, as mere nodes) and the relevant properties, such as spin, as presented.

However, it is philosophers of science, of course, who use (partial) set-structures to represent theories, their inter-relationships, both with each other and, heading downwards, with data structures etc., and moving up, with the families of mathematical structures into which theories can be embedded. Furthermore, these devices enable us to formalise and sharpen notions such as positive, negative and neutral analogies and allow us, of course, to draw on all the resources of set theory. Partial isomorphisms and homomorphisms can then be considered two of the various tools that philosophers can use in this representational activity. Thus at the meta-level where philosophers of science operate, it is (partial) set-structures that are doing the (meta-level representational) work.

So, I agree with Brading and Landry when it comes to the contextual determination of appropriate object-level structure (and hence appropriate structural ontology), where the context here is understood physically, rather than, say, culturally or sociologically. However, where we disagree is on the need for a meta-level representational unitary framework (provided by set-theory, category theory, whatever). Brading and Landry insist that it is 'shared structure' (group-theoretic, in the above case study) that does all the work but the work that is being done is 'physical' (!) work and while I agree that this is appropriate for physicists, philosophers are doing a different kind of work, that requires a different set of tools. To insist that this form of work should be dismissed would be a radical step that would fundamentally revise our conception of what the philosophy of science is all about. Let me now briefly recall the various modes of representation that can be called upon at the meta-level of the philosophy of science to represent the 'shared structures' that are deployed at the object-level of science itself.

1.11 Modes of Representation (at Meta-level) of Shared Structure

Perhaps the most well-known mode of representation in this context is the Ramsey sentence, obtained by replacing the theoretical terms of a theory with variables bound by existential quantifiers:

$$T(t_1, \cdots t_n; o_1, \cdots o_m) \rightarrow (\exists x_1), \cdots (\exists x_n) T(x_1, \cdots x_n; o_1, \cdots o_m) \qquad (1.1)$$

Philosophically what this amounts to is contentious, with different philosophers rediscovering it and putting the technique to different uses [9]. The relevant context of course is a form of the so-called 'syntactic' approach in which a theory is represented in terms of a set of axioms, from which various theorems can be deduced, and related to the propositions expressing the 'facts to be explained' by what Ramsey called a 'dictionary', taking the form of a series of definitions understood as equivalences [48]. The Ramsey sentence is alleged to capture the theory's structural (and empirical) content ([58]; for further details and concerns, see [54]) but as is well known it is widely held to be susceptible to a form of the so-called Newman problem: putting it bluntly and briefly, as long as the given theory is empirically adequate and has a model of the right cardinality, we can always find a system of relations definable over the relevant domain such that the Ramsey sentence is true. The claim then is that if structural realism uses the Ramsey sentence as her chosen representational mode, her realism will be trivialised.

I won't cover all the responses to the this argument, except to note that it depends on an extensional understanding of relations and as Melia and Saatsi point out, given the manner in which scientific theories are infused with modality, the realist—structural or otherwise—has good grounds for rejecting its premises and the Ramsey sentence itself [43]. Votsis and Frigg dismiss this response on the grounds that it is not clear how to motivate the kinds of modal operators envisaged [54] but the motivation seems clear and the Melia and Saatsi approach is surely the way to go, particularly given the long-standing emphasis on the modal nature of the structures in structural realism [19, 34].

And of course, there are well-known objections to the syntactic approach. I won't rehearse these here, nor the issue of its relationship to the so-called 'semantic' or set-theoretic approach, which represents theories in terms of set-theoretic models. As I have said, in his original paper introducing ontic structural realism, Ladyman urged the adoption of this approach on the grounds that it effectively wears its structural commitments on its formal sleeve [34]. Furthermore, if we recall 'the beasts of the field', the ability of the semantic approach to appropriately capture not only theory-data relationships, but also those that hold between theoretical and mathematical structures [7] and, of course, inter-theory relationships, appears to give it the advantage [12].

And it holds this advantage over the kind of category-theoretic approach advocated by Landry as well [37]. As briefly indicated by da Costa and French [12, p. 26], one could certainly consider representing theories in such terms but it's not clear what would be gained given the level of abstraction at which the relevant categories sit. Indeed, it might be suggested, rather crudely perhaps, that whereas category theory is extremely useful for expressing the relationships between different kinds of mathematical structures, what we tend to have in the case of scientific theories are the same kinds. And in particular, when it comes to the kinds of inter-theory relationships that are presented as a response to one of the principal motivations of structural realism, namely capturing the relevant commonalities in theory change, it is unclear whether category theory offers a better framework than the set-theoretic one.

Now this might be challenged in two ways. First of all, category theory might offer a useful meta-(meta-?!) framework for representing the inter-relationship between the two aspects of ontic structural realism arising from its twin motivations: on the one hand we have a focus on inter-theory relationships as just noted; on the other, we have the group-theoretic representation of objects and properties. One suggestion might be that category theory could offer an appropriate way of characterising the relationship within structural realism between these two aspects via the relationship between the categories 'Set' and 'Group'. Of course, the response might be that the relationship between the laws of a theory and the symmetries is already nicely captured set-theoretically, but nevertheless, the role of internal symmetries in this context might push us towards a more general category theoretic account.

The second would be to consider whether category theory offers a better framework for ontic structural realism because a category is characterised by its morphisms and not the relevant objects, with the latter regarded as secondary at best, or as definable in terms of, and consequently but more radically perhaps, reducible to, the morphisms going in and out. Thus category theory may seem to offer a way of representing the shift in focus from objects to structures that is central to OSR.

Certainly the set-theoretic representation appears inelegant at best in this regard. If we recall Cantor's original formulation, and its motivation, we can see that a commitment to objects appears to lie at the heart of the origins of the theory and even if we introduce novel formulations that capture the sense in which these objects might not be individuals, that commitment remains. This is not to say that there aren't ways of handling the structural realist's 'reconceptualisation' of objects within the set-theoretic framework [16, 19, 27]. We can perform what I have called the 'Poincaré manoeuvre': we begin with the standard presumption that theories are committed to objects, at least as the subjects of property instantiation; we then reconceptualise and, on the more 'radical' form of OSR eliminate, those objects in structural terms. Thus the putative objects come to be seen as merely stepping stones or heuristic devices to get us to the relevant structures. Given the initial presumption, it may seem natural to employ a set-theoretic representation, which includes the putative objects of course, but then we must insist that this be read 'semitically'; that is from right to left, so that, taking the simple formula:

$$\langle A, R \rangle \tag{1.2}$$

the relations R are understood as having ontologically priority over, and can be understood as constituting, the objects of the domain A.

Thus we seem to be faced with the following situation: the set-theoretic framework nicely captures the various inter-theory and maths-theory relationships that the structuralist will be interested in but has to be manoeuvred into accommodating the shift away from objects; whereas category theory has that shift 'built in' as it were, but operates at too high a level to straightforwardly capture the inter-theory relationships etc. In the spirit of a pragmatic and possibly pluralist approach to this issue of meta-level representation one option would be to follow a Suppesian line and suggest that when it comes to accommodating the structuralist response to the pessimistic meta-induction we adopt an 'external' characterisation of the relevant inter-relationships in set-theoretic terms, then shift to an 'internal' or ontological characterisation through category theory in order to capture the implications of modern physics for the notion of object.

One worry about moving to a category-theoretic representation is that the very abstract nature of this representation that makes it so powerful in mathematics may undermine its deployment in the service of structural realism. Thus Marquis notes that a standard objection to the use of category theory to underpin a form of structuralism with regard to mathematical objects is that regarded as abstract structures in this way, they cannot be appropriately located within the mathematical universe [42]. As he puts it, mathematical objects can be regarded as types that are tokened in different contexts. In set theory, the reference of mathematical objects is given directly and their identity pinned down via the axiom of extensionality. But

> [i]n a categorical framework, one always refers to a *token* of a type, and what the theory characterizes directly is the type, not the tokens. In this framework, one does not have to locate a type, but tokens of it are, at least in mathematics, epistemologically required. [42].

Marquis sees this as '... simply the reflection of the interaction between the abstract and the concrete in the epistemological sense (and not the ontological sense of these latter expressions.)' [42] but it is precisely in reflecting upon the nature of this 'interaction' that concerns arise. Consider the analogous situation when it comes to physical objects: in that case to move from the kinds presented to us by group theory to the concrete tokens that we observe, we attach to the relevant group an imprimitivity system that establishes the 'interaction' Marquis refers to [8, 19]. It is not clear what could serve the same function in the case of category theory, even granted the point that we need to be clear on what counts as a token in the context.

However, it may be that shifting to an *in re* interpretation can help illuminate things [46]. In suggesting that a category provides the schema for our discourse about mathematical structures, Landry urges a move away from the reification of structures and towards (again) shared structure of mathematical systems [38]. Category theory is thereby seen as a language in terms of which we can analyse systems that are structured, rather than as a 'meta-science' of structures. As such, it allows one to be an *in re* mathematical structuralist without being committed to categories-as-objects existing in some meta-linguistic structure [38]. Nevertheless, it presents an ineliminable framework that is 'prior in definition' to any particular system without being committed to the claim that mathematics is 'about' actual or possible objects and structures; in this latter sense, then, 'it is philosophy without either metaphysics or modality.' [38].

However, as Landry herself makes clear, importing such distinctions into the philosophy of science is problematic (Landry, see Chapter 2). I have also argued that the classification of '*ante rem*' vs. '*in re*' structuralism may not be appropriate in the context of OSR, where the world-as-structure is neither abstract, in the sense of being non-causal nor a system that is structured, in the sense that there is a system that is ontologically prior to the structure [19]. In particular, the pulling back from metaphysics and modality would be highly questionable in this context (as Landry herself acknowledges). As indicated above, it is in its presentation of putative 'objects' and their properties that group theory contributes to a metaphysics of them (what is spin? It is, in large part, the interweaving of relations that group theory presents to us) and category theory's contribution to such a metaphysics is attenuated by the comparatively higher level at which it operates. Thus, given the criticism of OSR that it invokes an abstract sense of structure that is inappropriate to the physical domain, one can see how concerns might arise. In particular, a category-theoretic reconceptualisation of physical objects in terms of the relevant morphisms 'in and out' may be too abstract to capture the relevant physical particularities, by virtue of the high-level nature of these morphisms. Perhaps, again, the best we can hope for is some kind of trade-off between the comparative and relevant power of different representational frameworks.

1.12 Conclusion: Presentation and Representation

There is certainly further work to be done here but the point I want to drive home is that we need more than a notion of shared structure. Without some formal framework, set-theoretic or otherwise, that can act as an appropriate mode of representation at the meta-level, our account of episodes such as the introduction of group theory into quantum mechanics would amount to nothing more than a meta-level positivistic recitation of the 'facts' at the level of practice. Any concern that the choice of a set theoretic representation of such an account would imply that set theory is *constitutive* of the notion of structure can be assuaged by insisting on the above distinction between levels and modes of representation. To reiterate: at the 'object' level of scientific practice, group theory was introduced and used to represent physical objects, their properties and the latter's relevant inter-relationships. This is the mode by which these objects are *presented* at this level. At the meta-level of the philosophy of science, there exists a variety of modes by which we can represent both this practice and our structural commitments. In deploying the semantic approach, or partial structures, there is no suggestion that, first of all, physicists themselves had such an approach in mind when they applied the mathematics that they did, or related the theories in the way they did (and Brading and Landry acknowledge that they are not implying that such a suggestion is being made); nor should this be taken to imply the view that the world is somehow, in some platonic sense, set-theoretical. The claim is merely that in order to appropriately represent the physicists' representation of the phenomena, the semantic approach offers a number of advantages to the philosopher of science, and in particular, for the structuralist by 'making manifest' the relevant structures.

Furthermore, as I have said, there is certainly a degree of context dependence here in the sense that the physical context 'reveals' and hence presents that aspect of the world-structure that is represented by group theory. And I agree on the significance of 'shared structure' in this sense (that is, at the object level) for the metaphysical presentation of the afore-mentioned objects and their properties. It is certainly this shared group-theoretic structure that is doing the work for the physicists at this level and not partial structures or anything of that kind (except maybe implicitly if one accepts set-theoretic reductionism). But I disagree that this is sufficient: at the meta-level where philosophers operate, it is (partial) set-structures that are doing the (meta-level representational) work (at least in the account I have offered). Within such an account, the structure is represented set-theoretically but the putative objects are presented and re-conceptualised (and hence metaphysically eliminated *qua* objects) via group-theory and it is the particularities of the latter's representations (in the technical sense) that reveal, represent and present to us the concrete features of the structure of the world.

Acknowledgments My thanks to Elaine Landry for organising the Banff workshop in 2007 where an early version of this paper was presented and to all the participants for helpful comments and criticisms; particular thanks go to Katherine Brading, Angelo Cei, Anjan Chakravartty, James Ladyman, Elaine Landry, Kerry McKenzie and Ioan Muntean for useful discussions on these and related topics. As always, the blame for any errors, infelicitous phrases or poor philosophy is entirely ~~theirs~~ mine. Finally, I would like to thank the Leverhulme Trust for the award of a Major Research Fellowship which enabled me to complete this work.

References

1. Bonolis, L. (2004) From the Rise of the Group Concept to the Stormy Onset of Group Theory in the New Quantum Mechanics: A Saga of the Invariant Characterization of Physical Objects, Events and Theories. *Rivista del Nuovo Cimento* **27**: 1–110.
2. ———— (2010) On Composite Systems: Newton, Descartes and the Law-Constitutive Approach. In P. Anstey and D. Jalobeanu (eds.), *Vanishing Matter and the Laws of Motion Descartes and Beyond* (pp. 130–152). London: Routledge.
3. ———— and E. Landry (2006) Scientific Structuralism: Presentation and Representation. *Philosophy of Science* **73**: 571–581.
4. Bueno, O. (1997) Empirical Adequacy: A Partial Structures Approach. *Studies in History and Philosophy of Science Part A* **28** (4): 585–610.
5. ———— (2000) Empiricism, Scientific Change and Mathematical Change. *Studies in History and Philosophy of Science* **31**: 269–296.
6. ———— and S. French (1999) Infestation or Pest Control: The Introduction of Group Theory into Quantum Mechanics. *Manuscrito* **22**: 37–86.
7. ————, ———— and J. Ladyman (2002) On Representing the Relationship Between the Mathematical and the Empirical. *Philosophy of Science* **69**: 452–473.
8. Castellani, E. (1993) Quantum Mechanics, Objects and Objectivity. In C. Garola and A. Rossi (eds.), *The Foundations of Quantum Mechanics—Historical Analysis and Open Questions* (pp. 105–114). Dordrecht: Kluwer.

9. Cei, A. and S. French (2006) Looking for Structure in All the Wrong Places: Ramsey Sentences, Multiple Realizability, and Structure. *Studies in History and Philosophy of Science* **37**: 633–655.
10. ———— and ———— (2009) On the Transposition of the Substantial into the Functional: Bringing Cassirer's Philosophy of Quantum Mechanics into the 21st Century. In M. Bitbol, P. Kerszberg and J. Petitot (eds.), *Constituting Objectivity, Transcendental Perspectives on Modern Physics* (pp. 95–115). Dordrecht: Springer.
11. ———— and ———— (forthcoming) Laws and Objects: A Neo-Kantian Template for the Structuralism Debate.
12. da Costa, N.C.A. and S. French (2003) *Science and Partial Truth: A Unitary Approach to Models and Scientific Reasoning*. Oxford: Oxford University Press.
13. Debs, T. and M. Redhead (2007) *Objectivity, Invariance, and Convention: Symmetry in Physical Science*. Cambridge: Harvard University Press.
14. Eddington, A. (1941) Discussion: Group Structure in Physical Science. *Mind* **50**: 268–279.
15. Falkenberg, B. (2007) *Particle Metaphysics: A Critical Account of Subatomic Reality*. Berlin: Springer.
16. French, S. (1999) Models and Mathematics in Physics: The Role of Group Theory. In J. Butterfield and C. Pagonis (eds.), *From Physics to Philosophy* (pp. 187–207). Cambridge: Cambridge University Press.
17. ———— (2000) The Reasonable Effectiveness of Mathematics: Partial Structures and the Application of Group Theory to Physics. *Synthèse* **125**: 103–120.
18. ———— (2003) Scribbling on the Blank Sheet: Eddington's Structuralist Conception of Objects. *Studies in History and Philosophy of Modern Physics* **34**: 227–259.
19. ———— (2006) Structure as a Weapon of the Realist. *Proceedings of the Aristotelian Society* **106**: 167–185
20. ———— (2007) The Limits of Structuralism. BSPS Presidential Address: http://www.oxfordjournals.org/society/bsps/french_talk.
21. ———— (2008) Symmetry, Invariance and Reference. In M. Frauchiger and W. K. Essler (eds.), *Representation, Evidence, and Justification: Themes from Suppes* (pp. 127–156). Lauener Library of Analytical Philosophy, Vol. 1. Frankfurt: Ontos-Verlag.
22. ———— (2010) Keeping Quiet on The Ontology of Models. *Synthèse* **172**: 231–249.
23. ———— (2010) The Interdependence of Structures, Objects and Dependence. *Synthèse* **175**, Supplement 1: 89–109.
24. ———— (forthcoming) One Trope or Two: How Much Metaphysics Should the Realist Take With Her Tea?
25. ———— (2011) Metaphysical Underdetermination: Why Worry? *Synthèse* **180**(2): 205–221.
26. ———— and J. Ladyman (2003) Remodelling Structural Realism: Quantum Physics and the Metaphysics of Structure, and The Dissolution of Objects: A Reply to Cao. *Synthèse* **136**: 31–56 and 73–77.
27. ———— and ———— (2011) In Defence of Ontic Structural Realism. In P. Bokulich and A. Bokulich (eds.), *Scientific Structuralism* (pp. 25–42). Boston Studies in the Philosophy of Science. Berlin: Springer.
28. ———— and J. Saatsi (2006) Realism about Structure: The Semantic View and Non-linguistic Representations. *Philosophy of Science* (Proceedings) **73**: 548–559.
29. ———— and P. Vickers (2011) Are There No Such Things as Theories? *British Journal for the Philosophy of Science* **62**.
30. Guay, A. and B. Hepburn (2009) Symmetry and Its Formalisms: Mathematical Aspects. *Philosophy of Science*, **76**: 160–178.
31. Ihmig, K.-N. (1999) Ernst Cassirer and the Structural Conception of Objects in Modern Science: The Importance of the Erlangen Program. *Science in Context* **12**(4): 513–529.
32. Kantorovich, A. (2003) The Priority of Internal Symmetries in Particle Physics. *Studies in History and Philosophy of Modern Physics* **34**: 651–675.

33. ———— (2009) Ontic Structuralism and the Symmetries of Particle Physics. *Journal for General Philosophy of Science* **40**(1): 73–84.
34. Ladyman, J. (1998) What is Structural Realism? *Studies in History and Philosophy of Science* **29**: 409–424.
35. ———— (2007) Structural Realism. *Stanford Encyclopaedia of Philosophy* http://plato.stanford.edu/entries/structural-realism/.
36. ———— and D. Ross (with D. Spurrett and J. Collier) (2007) *Every Thing Must Go: Metaphysics Naturalized.* Oxford: Oxford University Press.
37. Landry, E. (2006) Category Theory as a Framework for an *In Re* Interpretation of Mathematical Structuralism. In J. van Benthem, G. Heinzmann, M. Rebuschi and H. Visser (eds.), *The Age of Alternative Logics: Assessing Philosophy of Logic and Mathematics Today*. Dordrecht: Kluwer.
38. ———— (2007) Shared Structure Need Not Be Shared Set-structure. *Synthèse* **158**: 1–17.
39. Lyre, H. (2004) Holism and Structuralism in $U(1)$ Gauge Theory. *Studies in History and Philosophy of Modern Physics* **35**: 643–670.
40. MacKenzie, K. (forthcoming) The Fundamentality of Structure. Arguing Against Fundamentality forthcoming in Studies in History and Philosophy of Modern Physics.
41. Mackey, G.W. (1978) *Unitary Group Representations in Physics, Probability, and Number Theory*. The Benjamin/Cummings Pub. Co.
42. Marquis, J.-P. (2007) Category Theory. *Stanford Encyclopaedia of Philosophy* http://plato.stanford.edu/entries/category-theory/.
43. Melia, J. & J. Saatsi (2006) Ramseyfication and Theoretical Content. *The British Journal for the Philosophy of Science* **57**: 561–585.
44. Morrison, M. (2007) Spin: All is Not What it Seems. *Studies In History and Philosophy of Modern Physics* **38**(3): 529–557.
45. Nozick, R. (2003) *Invariances*. Cambridge: Harvard University Press.
46. Psillos, S. (2001) Is Structural Realism Possible? *Philosophy of Science (Supplement)* **68**: S13–S24.
47. Quine, W.V.O. (1951) Two Dogmas of Empiricism. *The Philosophical Review* **60**: 20–43.
48. Ramsey, F. (1978) Theories. In D.H. Mellor (ed.), *Foundations: Essays in Philosophy, Logic, Mathematics and Economics* (pp. 103–104). London: Routledge and Kegan Paul.
49. Redhead, M. L. G. (1975) Symmetry in Intertheory Relations. *Synthèse* **32**: 77–112.
50. Roberts, B. (2011) Group Structural Realism. *The British Journal for the Philosophy of Science* **62**(1): 47–69.
51. Saunders, S. (2003) Structural Realism, Again. *Synthèse* **136**: 127–133.
52. ———— (2003b) Physics and Leibniz's Principles. In K. Brading and E. Castellani (eds.), *Symmetries in Physics: Philosophical Reflections* (pp. 289–307). Cambridge: Cambridge University Press.
53. Suppes, P. (2000) Invariance, Symmetry and Meaning. *Foundations of Physics* **30**: 1569–1585.
54. Votsis, I. and R. Frigg (2011) Everything You Always Wanted to Know About Structural Realism But Were Afraid to Ask. *European Journal for Philosophy of Science* **1**(2): 227–276.
55. Weyl, H. (1931) *The Theory of Groups and Quantum Mechanics*. London: Methuen and Co.; English trans. 2nd Ed., 1928.
56. Wigner, E.P. (1937) On the Consequences of the Symmetry of the Nuclear Hamiltonian on the Spectroscopy of Nuclei. *Physical Review* **51**: 106–119.
57. ———— (1959) *Group Theory and Its Application to the Quantum Mechanics of Atomic Spectra*. New York: Academic.
58. Worrall, J. (2007) Miracles and Models: Why Reports of the Death of Structural Realism May Be Exaggerated. In A. O'Hare (ed.), *Philosophy of Science* (pp. 125–154). Royal Institute of Philosophy,vol. **61** (pp. 125–154). Cambridge: Cambridge University Press.

Chapter 2
Methodological Structural Realism

Elaine M. Landry

Scientists believe in the existence of electrons, bosons, fermions, fields, forces, space-time, etc.; they, unlike their philosophical realist counterparts, do not believe, however, in the existence of phenomena or noumena, observables or unobservables, detection or auxiliary properties,[1] etc. The aim of this paper is to carve out a naturalistic,[2] or *methodological*, structuralist account that serves to underpin scientists' belief in, for example, bosons and fermions *via* those structural properties or relations that are known by considering the shared structure between those models (both theoretical and data models) that are taken to *present* the content and structure of what we say about them as *kinds of objects*. With realism, the claim that such models *represent* the content and structure of what we say about *this* object as such a kind, resulting from the belief that it would be a miracle if the structure of the these kinds did not match some structure of the world. I will consider each aspect of this account in its turn.

Section 2.1 will consider what I mean by *methodological* with the aim of showing, *contra* French,[3] that as methodological structuralists our task is focused on the "object-level of scientific practice" and not on the "meta-level of the philosopher of science". Section 2.2 will investigate what I mean by *structural* with the aim of showing, *contra* Psillos,[4] that, as for the mathematical structuralist, for the scientific structuralist there is no distinction between nature and structure, or form and content. And so there is no philosophical position from which to talk about

---

[1] See [17], especially pp. 47–48.

[2] By "naturalistic", I mean only that claims of existence are to be made by scientists, not by philosophers, and too that the methods that give warrant to such claims are scientific, not philosophical. I do not also mean that the methods of philosophy *are* the methods of science. That is, I do believe that philosophers have the task of the analysis of the content and structure of what scientific theories express. But it is a task that, when we consider the expression of what we *say exists*, depends on the work of scientists.

[3] See Chapter 1.

[4] See [57].

E.M. Landry (✉)
Department of Philosophy, University of California, Davis, CA, USA
e-mail: emlandry@ucdavis.edu

E.M. Landry, D.P. Rickles (eds.), *Structural Realism*, The Western Ontario Series in Philosophy of Science 77, DOI 10.1007/978-94-007-2579-9_2,

objects as *individuals* that may have a *nature* as distinct from their structure. Finally, Section 2.3 will work-out what I mean by realism. My aim here is to use Worrall's,[5] suggestion, that NMA (the no miracles argument) only has its "carving" power when used with respect to particular *individual* theories, to argue for a *methodological structural realist* [MSR] position, whereby claims of the existence of electrons, bosons, fermions are made from *within* a mathematically framed, structurally presented, mature, empirically successful, scientific theory that shares the appropriate kind of structure with its predecessor and will share this structure with its successor.

2.1 Methodological

In this section I aim to provide an historical backdrop to my claim that debates about, or accounts of, scientific realism, need neither a metaphysical, logical nor mathematical *meta-linguistic* framework. The work of the *logical positivists* can be taken as bearing witness to what I mean by the claim that such accounts do not need a *metaphysical* meta-linguistic framework. It is a well-repeated fact that at least one of the goals of the *logical positivists* was the abandonment of metaphysics. From Russell to Wittgenstein to Carnap, the common thread weaving its way through their work was to replace metaphysics' role as a meta-linguistic framework *for what exists* with logic as a meta-linguistic framework for *what we say*. The idea here being that much of what we say is sullied, confused and obfuscated by metaphysical presumptions of what we can say about what exists. For example, by presuming that talk of substances, essences, causes, or Kantian categories, intuitions or Ideas, etc., is somehow or other constitutive or regulative of what we say, of what exists, or both.[6]

The turn underlying this shift from metaphysics to logic is what is standardly known as the linguistic, or semantic, turn. As Coffa characterizes this: the "semantic philosopher" is one, like Bolzano, who sees that, "the proper prolegomena to any further metaphysics was a study not of transcendental considerations but of what we say and its laws and that consequently the *prima philosophia* was not metaphysics or ontology but semantics" [20, p. 23]. That is, the semantic philosopher realizes that "between our subjective representations and the world of things we talk about, there is a third element, *what we say*" [20, p. 77], italics added.). The question at hand, however, was how can *representations*, borne out of what we say, be *objective*. The answer was to be given by a conceptual analysis focusing on the shared logical structure of what we say; giving rise to the slogan that, in some sense, "structure yields objectivity".[7]

---

[5] See Chapter 4.

[6] See [20], for an excellent overview and extensive account of these issues.

[7] For a thorough and insightful analysis of Carnap's belief that "structure yields objectivity", as intended and used in the *Aufbau*, see [62], especially chapter 2, section 2. There are, however, two notions of objectivity to be found in Carnap. As Richardson explains: "On the one hand, he constructs a notion of objectivity within the system of scientific concepts itself via the construction

This shift marks too a shift in the domain of our conceptual analysis; a shift from Kantian metaphysical, or psychological, analyses of concepts to logical analyses of the content and structure of what we say about the construction of concepts.[8] Building off of Frege's success[9] of using logic as a framework for the conceptual analysis of the content and structure of what we say in arithmetic, Russell, Wittgenstein and Carnap sought to use logic as a meta-linguistic framework for the analysis of the content and structure of the language in which we express what we say about the world. Again, this with the added aim of showing that philosophical analysis can get on just fine without Idealist metaphysics, Kantian psychologism,[10] or both! In any case, the sullied, confused and obfuscated "philosophical" meta-language of, perhaps psychologistically motivated, metaphysics is to be replaced by the pure, precise and rigorous meta-language of logic.

Yet there are two parts to this "abandonment of metaphysics", the first as detailed (but briefly) above is the logical part, the second, as should come as no surprise given that logical positivists are also known as *logical empiricists*, is the empirical part. That is, not only was this logical meta-linguistic framework intended to make rigorous our expression of the content and structure of what we say about concepts, but, in virtue of this, it was also to rigorize what, as empiricists, we say about the content and structure *of the world*. Under the influence of Wittgenstein's writing in the *Tractatus*, somehow or other, the *logical structure of language* was thought to match the *empirical structure of the world*, and vice versa.[11] Thus, we see why the appeal to structure, for the logical empiricist, is paramount.

---

of the intersubjective world of science (Sections 142–147). On the other hand, he also endorses the project of objectivity as pure logical structure through his notion of a 'purely structural definite description' (Sections 14–16).

[8] Note here Kant's distinction that philosophy deals with the *analysis* of concepts and mathematics with the *construction* of concepts. The logical positivists' aim, in contrast, was to show, as Frege had shown of mathematics, that philosophy deals with the analysis of the content and structure of what we say via logical construction of its concepts.

[9] See, especially, [24, 27].

[10] The "psychologistic interpretation" of Kant can be characterized as the view that the Categories, and the pure forms of space and time, are aspects of our make-up that arise from our psychological constitution. That is, while the forms of space and time arise from the constitution of our sensory faculty, the Categories arise from that of our intellect. In either case, they are the psychological (as opposed to logical and/or transcendental) conditions for the possibility of the objective *representation* of our subjective knowledge. See [24] for Frege's reaction to what he sees as Kant's psychologistic interpretation of arithmetic. The "logical interpretation", on the other hand, sees the Categories, and the forms of space and time, as logical requirements for knowledge. They are not grounded in the way we are constituted, but are taken to be features of our epistemic situation—conditions or rules that have to be adhered to if we are to gain knowledge. That is, they are the logical and/or transcendental conditions for the possibility of the objective *presentation* of our subjective knowledge. (See [29], for a reading of Kant along these, logical lines.)

[11] Note here that while this can be said of Carnap's work in, for example, *LSL* or the *Aufbau*, things change when we get to *ESO*. That is, in *ESO* it is not the logical structure of language per se but rather the structure of a "linguistic framework". This structure as determined by logical rules in the case of mathematical frameworks and by empirical rules in the case of scientific frameworks.

Let's now turn to consider two logical empiricists who appealed to structure in an attempt to detail how the logical structure of language could be said to match the empirical structure of the world. First we consider Russell. For Russell[12] we come to know the structure of the world in two ways; either directly via acquaintance or indirectly via description. Knowledge by acquaintance is somehow or other a purely empirical endeavor. It is through our sense experience that we come to have direct *percepts*, that is, direct perception of the way the world is. *To talk about this knowledge*, and thereby allow for the possibility of extending it, we must turn to knowledge by description. It is here that we find "structural" knowledge, or knowledge of the properties and relations of percepts; this is said to be knowledge of *concepts*. And yet to talk about such knowledge as, somehow or other, an objective expression of what we say *about the world*, we must presume that the structure of our percepts (those things we come to know directly by experience) matches the structure of our concepts (those things we come to know indirectly by description) and vice versa.[13] As Russell explains:

> I shall assume that the physical world, as it is independently of perception, can be known to have a certain structural similarity to the world of our percepts, but cannot be known to have any qualitative similarity. [64, p. 138].[14]

So Russell appealed to two presumptions: one meta-epistemological, that the structure of what we say about our percepts matches that of our concepts, and one meta-linguistic, that the notion of structure itself can be logically framed by some higher-order type theory, whereby theoretical terms are just contentless definite descriptions. Likewise Carnap[15] makes two presumptions, again, one meta-epistemological, that the observational/theoretical distinction can be expressed (though not defined) solely in terms of observables via correspondence rules and so even theoretical terms can be indirectly matched to the world, and one meta-linguistic, that the notion of shared structure itself can be logically framed, in terms of a "purely structural definite description",[16] by some higher order type-theory.

So far we have considered the logical and empirical aspects of Russell and Carnap; I now turn to rationally reconstruct the *naturalist* aspect of Carnap. I begin by claiming that the naturalist Carnap believes in the existence of electrons, etc., but, as we have seen, Carnap, as a logical empiricist, believes that he needs to speak about electrons *qua* either observables or unobservables. My question then is this:

---

[12] See [65].

[13] See Chapter 3 for a development and explanation of Russell's structuralism.

[14] This is an example of what Psillos [58] has called the "upward path" to ESR. As a means of resolving this tension between percepts and concepts and too of avoiding making this presumption, some philosophers advocating ESR, like Worrall, adopt a "downward approach", and yet stay within the "syntactic" view of scientific theories, by presenting a theory as a Ramsified sentence where the quantifiers that range over problematic unobservables are turned into existential quantifiers and so all we are left with is structure.

[15] See especially [9, 11, 12].

[16] See [12].

In what sense is Carnap a philosopher (logical positivist) and it what sense is he a naturalist? Carnap is a philosopher to the extent that he rejects metaphysics in favor of logic as the meta-linguistic framework for expressing the content and structure of what we say. This is the Carnap of *LSL*, the *Aufbau*, etc, where talk of "purely structural definite descriptions" and "unobservables" are terms of this "logic of language". They are not, however, part of the "language of science". As Richardson explains:

> Both of these projects can be viewed as methods of accounting for objectivity as a purely structural notion, but they differ in their notion of 'structure'. In the first case, the classical mathematical structure of physics is seen as the crucial objectifying structure for science. In the second, Carnap seeks to deploy the resources of logic to give a structural account for concepts that does not rely on this structure of the mathematized sciences but on the structure of type theory itself. ... The goal seems to be ... the elucidation of the unique role that the superadded mathematical structure of the world of physics provides for the question of the objectivity of science. [62, pp. 29, 76]

My aim will be to show that for the scientific realist only the first, "language of science", notion of "structure yields objectivity" is required. That is, I will argue that mathematical structure itself, and yet not just that of physics,[17] is the "crucial objectifying structure for science", so there is no need to "deploy the resources of logic" to "superadd" to mathematical structure to give a structural account of concepts, or to give, what I will explain as, a *structural presentation of kinds of objects*.

This is the sense in which I will rationally reconstruct Carnap of *ESO* as an naturalist, to the extent that he rejects "philosophical" frameworks in favor of "linguistic" frameworks, and to the extent that, as explained in the Richardson quote above, mathematical, as opposed to logical, structure can be taken as the objectifying structure for science. That is, one can see the work of the *Aufbau*, *LSL*, etc, as the attempt to show that the *metaphysical* carving of *nature* should be replaced by the *logical* carving of *language* with the aim of getting to a conceptual analysis of "purely structural definite descriptions" by meta-linguistically framing this language by type theory. But too one can see the work of *ESO* as the attempt to show that the logical carving of language (talk of observable or unobservables as terms of some syntactic theory *qua* a meta-linguistically expressed logical framework) should go hand in hand with the *scientific* carving of *language* (talk of electron as a term in some scientific theory *qua* a mathematically expressed linguistic framework). My aim then is to consider what work *mathematical* structure does at the "object-level" of *scientific practice* and, *contra* both Carnap and French,[18] for example, forego consideration of what work *logical or mathematical* structure does at the "meta-level" of *philosophy of science*.

---

[17] That is, I will not presume some type of reductionist programme whereby the sciences all reduce to physics and so physics become the bearer of structure and so of objectivity. I will, however, presume that the scientific theories that I am speaking of are mathematically expressible. For more of what I mean by this, see the last section where I characterise a scientific theory as, in Suppes' sense, a hierarchy of models.

[18] See Chapter 1.

It is in this scientific (or naturalistic) sense, then, that I intend to use the term methodological. That is, I hold that the claim that electrons exists takes place from *within* a mathematically expressed linguistic framework and is made in accordance with the methods, practices, rules, principles and laws that scientists use. Now Carnap famously, and problematically, thought that linguistic frameworks are distinguished to the extent to which the rules were logical, as in the case of mathematics, from those that were empirical or physical, as in the case of physics.[19] Insofar as this distinction was thought to rest on the analytic/synthetic distinction, this gave rise to his debate with Quine. Quine, in contrast, arguing for a pragmatic version of naturalistic holism.[20] Certainly, it is not my aim to enter this philosophical debate. And as a means of bypassing it altogether, I turn to the methods, rules and practices of science itself and note that insofar as science itself works both "top-down" from the theory and "bottom-up" from the data/data models its rules are both theoretical and empirical. I will have more to say about this distinction in the final section; for now, I merely offer it by way of a promissory note.

I've also said of this account that it is, in addition to being methodological, *minimal*. By minimal I mean that it requires no frame of frames. One begins with a scientific theory (*qua* a collection of mathematically framed models) as a linguistic framework; as a framework for expressing the content and structure of what we say about what exists, *without* the concern for the presentation of a meta-linguistic frame; either metaphysical, logical (type-theoretic or Ramsified type-theoretic) or mathematical (set-theoretic or category-theoretic). To this extent, I, like the logical positivists, merely presume that metaphysics has no role to play and yet, again, *contra* Carnap and French, nor do any other "philosophical" presumptions about the structure of language, the structure of scientific theories, or the structure of the world. Yet, like Carnap *qua* naturalist, I begin with a theory *qua* a linguistic framework and too with the methods, practices, rules, principles and laws, of the scientist, and aim to construct my philosophy of science from within this naturalistic stance.[21] As Maddy would express it, I take the stance of the "second-philosopher". But, unlike Maddy,[22] I feel no need to prefer a reading of Carnap as a "first-philosophy"

---

[19] See [9].

[20] For a more detailed account of Quine's semantic holism and its implications for Carnap's position see his "Two Dogmas of Empiricism". In particular, note his claim that "Carnap, Lewis, and others take a pragmatic stand on the question of choosing between language forms, scientific frameworks; but their pragmatism leaves off at the imagined boundary between the analytic and the synthetic. In repudiating such a boundary, I espouse a more thorough pragmatism. Each man is given a scientific heritage plus a continuing barrage of sensory stimulation; and the considerations which guide him in warping his scientific heritage to fit his continuing sensory promptings are, where rational, pragmatic" [61, p. 46].

[21] I intend to use the word 'stance' in a similar manner as it is used by van Fraassen [81] in his *Empirical Stance*. That is, as explained by Teller [77] as a "policy" with respect to the methodologies one adopts in order to generate "factual beliefs". See also [17, p. 17].

[22] See [52].

empiricist, over reading him as a "second-philosophy" *naturalist* (though admittedly a rationally reconstructed naturalist).

I have just said that I reject *any* meta-linguistic framework for scientific structuralism, including the mathematical frameworks of either set theory or category theory. Now it is true that when considering the various philosophical interpretations of *mathematical structuralism*, I have taken category theory as a meta-linguistic framework, and this too in the Carnapian sense.[23] But this is because I take mathematics as itself a language, so that a meta-linguistic analysis of the criterion of acceptability (consistency, coherence, satisfiability, etc) of a mathematical theory *qua* a structurally organized linguistic framework is needed.[24] But once a theory is so deemed as acceptable then what we say, from *within* a system that provides an interpretation of our theory, *presents what we say about what exists*. This is what allows us, in mathematics, to be a *semantic realist*[25]; a realist on the basis of the content and structure of *what we say* from within a linguistic framework.[26] A scientific theory, as expressed "semantically", by a hierarchy of mathematical models, by contrast, is a "linguistic framework" *with a purpose*; the purpose of *representing what exists in the world*, and not just *presenting what we say about what exists*. So its criterion of acceptability, be it empirical adequacy or approximate truth, must in some sense be tied to this aim. Another way of putting this point is that the scientific (as opposed to the mathematical) mark of acceptability must, in some sense, be given by some *correspondence* or *match* to the physical *world itself*, and not just to what we say about it. This, then, allows us to point to the problem of Carnap's "logical" construction of the world:

> In order to be a scientific realist, it is not necessary, or even reasonable, to believe that science is true; but it is necessary to believe it might be false. This is a test that Carnap's construction of the world fails miserably, for *it leaves us without the ability to distinguish between what science says and what is the case*. If Carnap were right, science could never be false. As science changes, Carnap would instruct us to reconstruct the world Carnap has succeeded in defining an anaemic sense of existence wonderfully fit to diffuse Platonism. It is as if he has taken abstract discourse as his target and then mechanically extended this result everywhere else. For those who think that electrons are somewhat more substantial than sets and that absolutely everything we say about the former may be wrong, Carnap had nothing to offer. [20, p. 239.]

---

[23] See [49].

[24] See [46].

[25] I use the term 'semantic realist' in a manner that is distinct from Psillos [59], who seems to take this to mean the position of the traditional scientific realist, that is, "[s]emantic realists are those, who following Feigl [23], argue that a full and just explication of *t*-discourse requires that *t*-terms have a factual reference to unobservable entities", p. 11. On my meaning, semantic realists are only committed to the acceptability of *what is said*, and this without relation to *what exists*. I also note a similarity of my use and Ladyman's in his [43], but I do not agree that the "semantic requirements" need be "cashed out in terms of a correspondence theory of truth", p. 158. For example, the mathematician may equally cash this out, as did Hilbert, in terms of relative consistency.

[26] See [48].

Before moving ahead, there are several loose ends that need now to be tied together. I have talked of philosophers following the linguistic turn as being concerned with the content and structure of what we say, and yet it appears, when considering the "syntactic" approaches of Russell and Carnap, and those who would have us Ramsify a theory, that we gain "logical" structure at the expense of "real world" content. To make my point, I focus here on Carnap. Carnap viewed an empirical theory "syntactically" as a *partially interpreted logical calculus*; that is, as a set of sentences (expressible in first-order logic and meta-linguistically framed by higher-order type theory), some of which, the observable sentences, were interpreted and some of which, the unobservable sentences, were only interpreted via "correspondence rules", which serve to link unobservable terms with observable terms.[27]

Again, it is not my aim to discuss problems associated with this view, but to point out that there are two assumptions typically attributed to logical positivists/empiricists: a "syntactic" view of the structure of scientific theories and a distinction between, or rather collapse of, content and structure, wherein the content of the sentences of a theory are expressed by their logical form or structure. This view is held, or so it is claimed, to be in contrast to those holding the "semantic" view of the structure of scientific theories, where a theory is thought of as a collection of models, so that the content of a theory (insofar as content is thought to be located in, or given by, an interpretation) comes with its structure. Interestingly both of these views find their history in the works of Hilbert.[28] Indeed, Carnap, tells us he is taking his view of theories, as axiom or postulate systems, from Hilbert's view of geometry.[29] And so is to Hilbert's view of geometry, and its relation to structuralism, both mathematical and scientific, that I now turn to consider.

2.2 Structural

In this section, I will consider the details and legacy of the Hilbertian view for the purpose of characterizing what I mean by 'structural', and I will use this to further investigate the on-going structural realist nature/structure debate (see Psillos[30]) by

[27] See [9].

[28] But too, as we will see, one must be careful with these distinctions; Hilbert was no formalist in either the "uninterpreted calculus" or the "mathematics is a game of meaningless symbols" sense! See [53], especially, pp. 162–164. And see [22] for standard readings and references, particularly, see the debate between Hilbert and Brouwer.

[29] See [9], especially, pp. 323–325, where Carnap says that while, like Hilbert, he will take a theory as an axiomatic, or postulate, system, that, in contrast to Hilbert's view of a mathematical theory as an uninterpreted axiom system, a calculus whose terms (like 'point', 'line' and 'plane') are uninterpreted until they are logically and fully interpreted by being *explicitly* defined, for a physical theory "in order to connect the uninterpreted terms (like 'electron') with the observable phenomena, we must have [correspondence] rules for establishing the connection (with something in the physical world)", p. 324.

[30] See [57, 58, 60].

locating it in the mathematical content/form debate. My aim is to show that these debates are a red herring for both the mathematical[31] and scientific structuralist, for whom, simply, there is no such distinction. As noted, Carnap took theories in general to be uninterpreted calculi, with empirical theories being partially interpreted by correspondence rules. This view of theories is no doubt tied both to his reading of Hilbert as in line with a formalist/logicist view of mathematical theories and to Carnap's own "syntactic" view of logic. That is, for Carnap, logic itself is taken "syntactically" as a *calculus*, so, in so far as the concepts of mathematical theories are logically definable, they are uninterpreted calculi. Lest, however, one be tempted to see this as the only alternative, I point out that for Frege logic came with content; it was a *language* about everything thinkable and so ranged over a fixed interpretation that set its truth conditions.[32] Indeed, Frege famously argued against the "syntactic", Boolean, view that logic was a calculus.[33] My aim is to now show that it is between these two poles of logic as an interpreted language, ranging over a *fixed* domain, and logic as an uninterpreted calculus, ranging over a *variable* domain, that we are best to situate Hilbert's view.

Again we turn to another of Frege's famous fights to pull out these distinctions. A theory for Frege must begin with *explicit definitions* of concepts, the axioms are then about these defined concepts (or the objects that fall under them) and so can either be true or false so long as they express the content of a true thought about such concepts, that is, so long as they refer to the conditions for its truth. For example, in geometry, one must begin with the *explicit definitions* of 'point', 'line' and 'plane' and then the axioms are such that they are truths about these concepts, so that consistency follows from the truth of the axioms. For Hilbert, in contrast, the direction is reversed; one starts with axioms, themselves, as *implicit definitions* of the concepts and modulo their relative consistency, claims of existence (of the objects that fall under these concepts) and the truth of the axioms follow.[34] That is, if one can give a relative consistency proof by interpreting the axioms in say another "acceptable" theory, like finitary arithmetic, then one is licensed to say that the axioms are true and points exits. Thus we note that, for Hilbert, the meaning/content of concepts is fixed over *variable domains*[35]; it is not, like it is for Frege, that the domain is fixed by a singular domain of thoughts, or by domain of "everything thinkable".

The point of Hilbert's "axiomatic approach" is that once the consistency of the axioms has been established, any interpretation can fix the meanings of the concepts

---

[31] See [50].

[32] See [35] for an excellent overview of the logic as language, logic as calculus debate and history.

[33] See [25, 26].

[34] For a hint that Carnap was confused about these distinctions see his claim, in [9, p. 324] that "[i]f we wish to give an interpretation of a term in a mathematical axiom system, we can do it by giving a definition in logic. Consider, for example, the term 'number' as it is used in Peano's axiom system. We can define it in logical terms, by the Frege-Russell method, for example. In this way the concept 'number' acquires a complete, explicit definition of the basis of pure logic".

[35] See [36, 37].

implicitly defined by them. Hence, Hilbert's famous quip that 'point', 'line' and 'plane' can be variably interpreted by even such terms as 'tables', 'chairs' and 'beer mugs'. So, *contra* Carnap's reading, it makes no sense to characterize Hilbert's view of axioms systems, or theories, either in a formalist sense, i.e., as an uninterpreted calculus of meaningless marks brought together through formal "syntactic" rules,[36] or a in logicist sense, i.e., wherein the concepts are explicitly defined on the basis of logic.[37] Thus, according to the Hilbertian view, it is not that theories come *without* interpretations it is that they come without *fixed* interpretations.[38]

Now one must note that for Frege, Hilbert and Carnap the differences between the "syntactic" and "semantic" views of both logic and theories are but just emerging.[39] Proof theory and model theory are just beginning to be rigorously considered, as too are the meta-mathematical relations between them. The significance of this is especially important for understanding the meta-mathematical and meta-logical work of both Hilbert and Carnap.[40] Indeed, it is just these meta-mathematical considerations that, in light of Gödel's results, motivate Carnap to adopt Tarski's logical framework,[41] with its "semantic" account of truth, whereby he could come to express, within a single framework, the needed Fregean distinction between the content of a statement and its assertion [20, p. 373].

In any case, regardless of how we interpret the history of their developing views, what is important, is that neither Frege nor Hilbert thought of theories "syntactically" as empty shells of form waiting for content. What was at issue is whether content fixes form or whether form fixes content, i.e., whether, like Frege, one begins with concepts whose content is fixed by *explicit definitions* expressed over a *fixed* interpretation, and then constructs axioms such that they assert truths. Or whether, like Hilbert, one begins with (relatively) consistent axioms systems and *implicitly defines* concepts whose content is fixed over *variable* interpretations. In any either case, one never finds content without form or form without content.

---

[36] Note here Hilbert's quip in reference to his response to Brouwer's charge of formalism: "And in many other instances one-sided prejudices and slogans are cheerfully introduced into the fray. I have already discussed the reproach of formalism in earlier essays. Formulae are a necessary aid to the logical investigation. To be sure, their use demands precise mental labour, and makes empty twaddle impossible." (Hilbert [1931a] in [22, p. 1156]).

[37] For a more detailed discussion of this point see [80], especially, p. 410, where Toretti says: "if philosophers had properly assimilated Hilbert, the "received view" of theories would not have made it to the twentieth century. But its modern advocates, notably the logical empiricists did not understand it in the manner explained above [wherein an axiom system is taken as a system of conditions for what might be called a relational structure [2] and not as a system of statements about a subject matter] and so were able to defend the "received view" in good conscience".

[38] Indeed, claims of existence and truth, in so far as they require a relatively consistency proof, depend crucially on there being *at least one* interpretation.

[39] See [41] for an excellent overview of the history of the development of logic.

[40] See [20].

[41] As Coffa notes: "The Hilbertian temptations to which Carnap was ready to succumb were bursting out of his Russellian framework" [20, p. 280].

Having made this point, it is here that our structuralist story can begin. The *mathematical structuralist* builds from its Hilbertian heritage[42] and too aims to keep its contrast with the Fregean.[43] That is, in light of the work of Benacerraf [1], it begins by rejecting the idea that numbers, for example, can be countenanced as "Fregean objects" *qua* individuals, that is, as objects that fall under concepts that, themselves, are explicitly defined, and so are individuated by their properties as fixed by some logical (or set-theoretic) interpretation.[44] The basic argument advanced by Benacerraf is that this "logical" approach cannot determine which (set-theoretic) interpretation is the "right" one to be used to "fix" the meaning of the concept number. For example, for Johnny's set-theoretic interpretation, numbers are explicitly defined in terms of Zermelo numerals, $3 \in 17$ and, yet, in Ernie's, numbers are explicitly defined in terms of von Neumann ordinals, $2 \notin 3$. Thus, one is bound to conclude that if numbers are "Fregean objects" *qua* individuals, that can be individuated or identified by their properties as fixed by logical/set-theoretic definitions alone, then numbers cannot be objects. It is in the light of this conclusion that one is urged by Benacerraf to adopt the "formist" approach, wherein one characterizes numbers only in terms of their shared "formal" features.

This is not, however, full-blown mathematical structuralism. The contemporary mathematical structuralist goes one step further than Benacerraf; he rejects the idea that objects are "Fregean objects", i.e., objects *qua individuals*, and yet, in so doing, recovers the notion of an object as nothing other than "a position in a structure". Just as, for example, there is no more to 'point', 'line' and 'plane' than the properties and relations that are fixed by the axioms, there is nothing more to '3' as an object, than the properties and relations that are fixed by the axioms. The mathematical structuralist like Hilbert, and unlike Frege, holds that there is no fixed domain of interpretation, in fact, one fixes the content of concepts across *various* and *variable* domains of interpretation. This with the result that an object is but "characterized up to isomorphism" as a position in *any* or *all* systems[45] that satisfy the axioms that characterise the concept number. So, for the mathematical structuralist, the Peano axioms, for example, are about natural numbers *qua any* or *all* systems that satisfy the axioms. In [8], we expressed this by claiming that the theory of natural numbers, as given by the Peano axioms, *presents* us with numbers as *kinds of objects*, as opposed to it *representing* numbers as "Fregean objects" *qua individuals*. The theory then is not about *the* numbers, it is equally about the Zermelo numerals, the von Neumann ordinals or, as Hilbert noted, a system of strokes on a page, e.g., I, II, III.

Now, for some, there is still the question of whether, at the *abstract level* of consideration of mathematical structuralism, form fixes variable content or variable

[42] See [80], fn. 15, for mention too of the Hilbertian heritage of *scientific* structuralism in the works of, for example, Minkowski, Weyl and von Neumann.

[43] See [46, 51, 71].

[44] Again, I point the reader to [9].

[45] See [56] for a discussion of the differences between mathematical structuralists' positions on that use *any* versus those that use *all*. For my purposes I will gloss over this distinction.

content fixes form. Those holding the latter are *ante rem* structuralists, those holding the former are *in re*. The *ante rem* mathematical structuralist, Shapiro [73] for example, holds that *actual structures* exist over and above any or all systems that have structure and that objects are thus fixed as "places" in a structure, so that this abstract structure serves to fix an interpretation in the Fregean sense. Thus, the *ante rem* mathematical structuralist speaks Platonistically of systems *exemplifying* the structure. The *in re* mathematical structuralist, Hellman [40] for example, eschews talk of abstract structures over and above systems that have a structure, so that objects, as positions in this system, or any *possible* system that has the same structure, are thus fixed variously. As an *in re* mathematical structuralist, Hellman, speaks modally and nominalistically of systems *having* a structure.

In either case, both the *in re* and *ante rem* structuralist take objects to be nothing but positions in an abstract structure or possible system, i.e., they have no other identity conditions, and so no other relevant properties and relations, than those specified by the axioms that are taken to characterize their structure. The *raison d'être* of mathematical structuralism, then, is that it can forego traditional debates about the independent existence, and/or absolute criterion of identity or individuation, of mathematical objects in favor of discussions about objects *qua* positions in a structure or system, whereby objects are *fully characterized* by their shared structure, i.e., are characterized, "up to isomorphism", by their structure as fixed by the axioms they satisfy.

For example, for the *ante rem* structuralist the natural number 3 is nothing but a position in a structure that satisfies the Peano axioms, where 'structure' means an actual freestanding ontological entity and 'position' means a "bare position" or "place" in this structure. For the modal-theoretic *in re* structuralist, the number 3 is likewise nothing but a position in a system that satisfies the Peano axioms, where 'system' is modally considered to mean *any* or *all possible* systems that have the same structure. In any case, the only "objects" that the *ante rem* structuralist is committed to are actual structures and the only "objects" that the *in re* structuralist is committed to are possible systems. Now, both of these commitments are problematic; that is, *ante rem* structuralism commits us to the actual (metaphysical) existence of structures and *in re* structuralism commits us to the possible (modal) existence of systems.[46] Indeed, what I have been at pains to show is that a category-theoretic version of *in re* mathematical structuralism, avoids these problems, by presenting a category as a *schema* for what we say about structured systems.[47]

In any case, *none* of these *mathematical structuralist* positions commit us to the existence of "places" or "positions", or to the "things" that "fill them", *qua* objects as *individuals* that have a "nature" independent of their "structure", and so have "other" properties (properties relating to their "nature" or "content") independent of their structural, properties (properties relating to their "structure" or "form"). So, it is simply an error to draw, as Psillos does, from the claim that mathematical

---

[46] See [51].

[47] See [46, 51].

structuralism "is not revisionary of the underlying ontology of objects with properties and relations", the "intermediate moral" that "structures need objects [as picked out from the "elements of a domain" or from an "ontology of individuals"]. This holds for both *ante rem* and *in re* structuralism" [57, p. 6]. As detailed, from the vantage point of both its Hilbertian heritage and its beginnings in the work of Benacerraf, the entire purpose of mathematical structuralism is to be "revisionary of ontology" in just this sense!

Both the *ante rem* and *in re* positions are committed to the view that there *is nothing more* to an object than its *structural properties*, i.e., the properties it has it virtue of the relations it bears to other objects as determined by the axioms. This holds whether objects are construed *ante rem*, as places in a structure, or *in re*, as positions in any or all systems that have the same structure. Both the *ante rem* and *in re* positions, then, are committed to the view that objects are characterized *solely* by their structural properties. Thus, *contra* [57, p. 5], no appeal to "objects" *qua* individuals, again, possibly having other, non-structural, properties is needed. Nor, indeed, would such an appeal make any sense!! And too no appeal to the "nature" or "content", causal or otherwise, of objects over and above their "structure" or "form" [57, p. 6] makes any sense. This point speaks as much against the Poincaré/Worrall [85] claim that what the Fresnel light example shows is that we can know the "structure" of light whilst remaining agnostic (or atheistic) about the "nature" or "cause" of light, as it does against Psillos' claim that "by leaving out entities we leave out structure as well" [58, p. 7]. And equally, is pointed at Psillos' various attempts to argue that since nature and structure "form a continuum" [60] and since causality may well pertain to the "nature" of the object, the scientific structuralist cannot be quite sure, as Hawthorne's [38] *causal structuralism* suggests, that scientific presentations of the structure of the world capture all that can be said about its "causal structure" [57, p. 12].

I pause here to respond in more detail to Psillos by drawing yet another lesson from mathematical structuralism. Does structuralism, either mathematical or scientific, commit us to the view, as Psillos seems to claim it does, that if there are "other" properties of these structurally presented objects *qua* positions, *then* these "other" properties might well be fixed by the "nature" of an object as an individual, as opposed to their being fixed by the "structure" of an object as a position? As I will next show, the answer is: no. Note that the consequent of this conditional is what supposedly allows for Psillos' further claim that it is just these "other" properties, that arise from the "nature" of an object *qua* a non-structurally presented "filler" of a position, that accounts for its identity and hence provide for the conditions of its individuation.[48] And, finally, note that it is what also allows for the conclusion that it might well be that it is such objects *qua* individuals that provide the locus for what

[48] See Psillos [57, p. 5] where he says, "According to *in re* structuralism there are no extra objects [like the *ante rem*'s "places"] which 'fill' the structure. It's then obvious that the objects that 'fill' the *in re* structures have more properties than those determined by their interrelationship in the structure. They are given, and acquire their identity, independently of the abstract structure they might be taken to exemplify".

bears the causal/nomological properties and relations that science seeks to capture. It is to these claims and conclusions that I now turn.

We can simply and immediately undercut Psillos' claims and conclusions by showing that these "other" properties need not be taken as "non-structural", unless one presumes, of course, that there is just *this* system of objects that *fixes* the domain of interpretation. As we have seen, in one set-theoretic system having the natural number structure, $3 \in 17$, and, yet, in another $3 \notin 17$. This does not mean, however, that the property $\in 17$, while being an "other" property of 3 *qua* a natural number, is a non-structural property. It means that the property $\in 17$ is not a "necessary" property of 3 *qua* a position in *any* or *all* systems *having a natural number structure*. Does this mean that this "other" property is fixed by, and so "necessary" in virtue of, the "nature" of 3 as opposed to its "structure". No. For example, it is fixed by the *set-structure of the Zermelo cardinals*, and so it is a structural property, which is necessary in virtue of the structure of this system that has the natural number structure.

One can only get to the conclusion that this property is fixed by the "nature" of 3, if *this* system is held to be *the* system that *fixes* the domain of interpretation, so that 3 *qua* an individual bears the "necessary" properties via its identity conditions. But as Benacerraf has shown us, forgoing, claims of logicism/set-theoretic foundationalism, we are thus, as mathematical structuralists, bound to conclude that all there is of 3, or all we know of 3, is that it is a position in *any* or *all* systems that have the same structure as characterized by the Peano axioms. In either case, the result is for the mathematical structuralist is the same: all "necessary" properties are, and must be, structural properties of objects *qua* positions or, to put it otherwise, no "necessary" properties can be fixed by the "nature" of an object *qua* an individual. To think otherwise, is to forgo mathematical structuralism.

Why is any of this important for the *scientific structuralist*? Because it speaks directly to the continued confusion of the nature/structure debate that one finds in the structural realism literature. Psillos tells us that in adopting the *in re* structuralist position we still leave open the possibility that there are "other" properties of objects, this licensing the additional claim the these "other" properties may well be properties pertaining to the "nature" of objects. So that one can never be quite sure, if the causal nexus is found in these "other" properties of the objects *qua* individuals, that structural properties and relations capture causal/nomonological properties and relations. The supposed result is that, as *in re* scientific structural realists, we cannot claim that all we know is structure (as the advocate of ESR would claim), or all there is, is structure (as the advocate of OSR would claim). My response is that the *in re* structuralist position forestalls neither position and so is a mere red herring. To see this, simply take 'necessity', as analysed above in purely structural terms, as a minimal component of what is meant by either causal/nomological.

Where then does Psillos go wrong? Simply, he confuses the Fregean and the Hilbertian views. That is, he assumes rightly, as does the Hilbertian, that

> [m]athematicians define and study all sorts of structures and any structure, defined implicitly by a set of axioms, will do

and again, he perhaps rightly assumes that

> things are more complicated when it comes to physical systems For finding the structure of a natural system in an a posteriori (empirical) enterprise. Its structure is in re. And it is a natural structure in the sense that it captures the natural (causal/nomological) relations among the objects of the system. It is *the* structure that delimits a certain domain as possessing causal unity' [57, p. 6].

But he is wrong, as is the Fregean, to assume that, as a *scientific structuralist*, one must *begin with* a "domain of elements" or "ontology of individuals" [58, p. 10], that *fixes* the interpretation so that

> it [the causal unity] is grounded on the causal relations among the elements of the domain, this so that "the phenomena are able to give 'content' to a structure precisely because they are *not* themselves structured". [57, p. 7].

This would be analogous to our logicist/set-theoretic foundationalist claiming that one must *begin with* a domain of logical/set-theoretic elements that *fixes* the interpretation of the concept number, whereby the "nature" of numbers provides for those identity conditions with would allows us to individuate them as "Fregean objects". But again, this is precisely what Benacerraf argued against, and too is what no mathematical structuralist following in his footsteps would accept. So to Psillos' claim that

> [u]nless we buy into some problematic metaphysical thesis which somehow 'constructs' the individuals out of relations, the world we live in (and science care about) is made of individuals, properties and their relations.

I reply that this last "thesis" is both decidedly non-structuralist and is a more "problematic metaphysical thesis", in so far, for example, as it opens the door to talk of "quiddities" [57, p. 12], than is the scientific structuralist's claim that objects are nothing other than positions in any or all systems that have the same structure.

2.3 Realism

So now on to the details of what I mean by *scientific structuralism*, with the aim of arguing for the position I have called *methodological structural realism*. In what follows, I will adopt an *in re* approach to scientific structuralism; I will talk of objects as *positions* in *physical systems* and of systems *having* a structure. Again, borrowing from my work with Brading, [8], and taking my lead from the above account of mathematical structuralism, I will presume that scientific theories *present* us with objects *qua kinds of objects*, as opposed to presuming that they *represent* objects *qua particulars*.[49] Recalling then my methodological stance, I take it that all that is needed for the expression of the content and structure of what we say about what exists occurs at the *mathematical level*. That is, no meta-linguistic framework,

[49] Here I point the reader to Chapter 5 and Brading's [6] argument that particulars need not, and should not, be taken as individuals.

either syntactic (as in Worrall's Ramsified-sentence approach) or semantic (as in Suppes' set-theoretic, Bourbaki,[50] approach or French's partial structures approach) is needed. This is what I mean when I say that that there is no semantic/syntactic distinction that matters for the scientific structuralist.[51] It certainly matters for the logician, and it might matter for the logical positivist who aimed to meta-linguistically formally frame what we say about *the* structure of scientific theories *qua* uninterpreted calculi, but from my *methodological stance*, this is rejected as "philosophy first". Thus, it is by the same light got by adopting a methodological stance, and for the same reasons, that I rejected Carnap's, or indeed any Ramsified meta-logical, type-theoretic account of the meta-linguistic structure that is supposed as underlying the "syntactic view" of scientific theories, that I reject[52] Suppes' or French's meta-mathematical, Bourbaki inspired, set-theoretical account of the meta-lingusitic structure that is supposed as underlying the "semantic view".

Again, instead of offering up a meta-linguistic account of what is meant by structure, I place my focus, as did Carnap in *ESO*, on what can be said of the match between what we say and what exists from *within* some *mathematically expressed* scientific theory *qua* a "linguistic framework" *and* this as determined by the methods, rules, etc., as employed by scientific practice. Further, as explained above, as someone who adopts a Hilbertian position, I take it that there is no distinction to be had between either content and form, or nature and structure. Finally, as someone wishing to remain true to Benacerraf's structuralist spirit, I take it that there is no meta-ontological category of objects as individuals, as entities, etc., that bear "other", non-structural, or causal, properties.

Moreover, as an *in re* scientific structuralist, I hold that there is no *abstract* notion of structure, either meta-linguistic or metaphysical, over and above *particular* systems that have a structure. Objects *qua* positions in any and all systems that have the same structure are thus *presented* by theories as *kinds of objects*. Finally, by 'model' I mean a *physical system*, that is, a system that *has* a structure in virtue of providing a *physical interpretation* for some mathematical or scientific theory, or for some empirical data. It is in this sense, then, that I consider a scientific theory as, in the sense of Suppes, a hierarchy of models ranging from theoretical models to data models. And, in so far as we can connect theoretical models to data models, I take *a* scientific theory, as a hierarchy of models, linguistically framed by some mathematical theory, to *present* the content and structure of what we *say* about what exists. Yet, too, *without*, as does French, aiming, in addition, to meta-linguistically

[50] See [46] and [80] pp. 412–413 for details of Suppes' use of Bourbaki's set-theoretic notion of structure.

[51] In fact Torretti [80] chooses to use the 'structuralist view' instead of the 'semantic view' for just this reason; claiming that "this appellation can only make sense at a time when there are people [like logical empiricists] who are willing to countenance a conception of theories that is not semantic. In saner times, the term will not pick out a specific concept of theories", p. 412, fn. 14. I too, for this reason and to remain true to its Hilbertian beginnings, will use the 'structuralist' over the 'semantic' view.

[52] See [46].

present the structure of scientific theories themselves and, in so doing, to *represent the structure of the world.*

The structuralist view of scientific theories, which characterizes a theory, in Suppes' sense, as a hierarchy of models that *share structure*, is put to use to account for: (a) the *applicability* of a mathematical theory to a physical theory,[53] e.g., the applicability of Hilbert-space formalism and of group-theoretic structure to quantum mechanics, (b) *the structure* of a scientific theory, e.g., the group-theoretic structure of a quantum mechanical theory, (c) the *applicability* of a physical theory to a set of empirical data, e.g., the applicability of the theory of quantum mechanics to a set of empirical data of, say, a particle's mass, spin, charge, etc, and (d) the *structural realist's* claim that the structure of the data model matches the structure of the world, e.g., the claim that all there is, or all we know, of quantum particles, is their Lie-group-structure.[54] These accounts are achieved by pointing to some morphism[55] that captures the relation of *shared structure* between (a) the model of *a* mathematical theory and the models of *a* physical theory, (b) the models of a physical theory, (c) the models of *a* physical theory and the models of the data, and (d) the models of successive, mature, empirically successful, physical theories.[56]

Recall now that the scientific structuralist's task, whether he is a scientific realist or not, is to use accounts of shared structure to talk about the content and structure of what we say about *what exists*. That is, as Suppes has pointed out [75, 76], scientific

---

[53] Again, my scientific structuralism is about the structure of a *particular* theory, for example, about the structure of *a* quantum mechanical theory. It is not about *the* structure of scientific theor*ies*. As noted, this marks a significant departure for the scientific structuralism of Suppes, French, etc.

[54] These are all case of *vertical* applicability. Of course there are examples, of *horizontal* applicability, for example when a theory from one domain is applied to another or when an independent and autonomous model "mediates" between the theoretical and the data model. I take these cases as equally important for the structuralist story, but for the sake of brevity I leave consideration of these examples out. See Chapter 1 [4] for more on horizontal applicability, and [55] for a detailed discussion of the role of, and problems associated with, the use of mediating models.

[55] Suppes uses isomorphism, French uses partial isomorphism, partial homomorphism, etc., each found at the "meta-level" of philosophical analysis. I have argued [47], that it is best to see the "appropriate kind of morphism" as fixed at the "object level" of the practice of science, i.e., as fixed with respect to some particular kind of theory and within some context, wherein the methods, rules, etc., appealed to in that context fix the kind of morphism by fixing the domain of interpretation. As I then claimed "[f]or example, in the context of speaking about the shared structure of systems structured by space-time theories the appropriate kind of morphism is a diffeomorphism, and this regardless of what a diffeomorphism *is*, i.e., regardless of whether it is a function between set-theoretic elements or an arrow between category—theoretic objects. Note also that if we narrow this context we must likewise narrow the kind of morphism. For example, while for generally relativistic theories the morphism between the dynamically possible models will be any diffeomorphism, for special relativistic theories the morphism between models will be a restricted kind of diffeomorphism called a Poincaré transformation and for Newtonian Mechanics it will be another restricted kind diffeomorphism called a Galilean transformation. The groups of Poincaré and Galilean transformations being subgroups of the diffeomorphism group.", p. 3, fn.3.

[56] For another example, see [6], which characterizes these uses and related appeals to shared structure for Newton's theory of gravitation.

theorizing consists of "a hierarchy of theories and their models" [75, p.255] that bridges the gap between the high level theory and the lower level "phenomena" that the theory is intended to *be about*. For Suppes, however, two things are required to connect the high level theory to the phenomena: an *experimental theory of the data* and an *empirical theory of the phenomena*. Suppes [74–76] details the evaluative criteria of theories of experimental design and of *ceteris paribus* conditions that go into the construction of the experimental theory of the data. But he is clear that, since there are, strictly speaking, no models of these theories, one can only characterize the experimental theory of the data by the collection of its data models; and so one's formal analysis must *begin with* models of data. To further connect the data models to the phenomena one must establish that they share structure. But, as Suppes notes, without an (empirical) theory of the phenomena, one cannot speak of "the structure of the phenomena". As a consequence, Suppes remains silent on the question of how we can suppose that models of data share structure with, and so apply to, the phenomena.

It is here, then, that we [8] noted three options in accounting for applicability of data models to the phenomena in terms of shared structure: (i) from a *methodological* stance, we may forgo talk of the structure of the phenomena and simply begin with structured data, that is, with data models, which themselves are taken to structure *what we say* about the world; (ii) from an *empirical* stance we may say that what structures the phenomena into data models is the high level theory[57]; and finally, (iii) from a *realist* stance[58] we may say that what structures the phenomena is the world.[59] In any case, regardless of the stance that one adopts, data models

---

[57] Suppes, for example, has it that what structures the phenomena into data models is the models of the experiment which themselves appeal to the high level theory. More for formally, for example, as [8, p. 578], noted "van Fraassen, as a *structural empiricist* suggests that we simply *identify* the phenomena with the data models: "the data model... is, as it were, a secondary phenomenon created in the laboratory that becomes the primary phenomenon to be saved by the theory" [81, p. 252]. In this way, the step from presentation to representation is made almost trivially: the data models act as the "phenomena to be saved" and so all we need to connect the theory to data models *qua* "the phenomena" is a guarantee of their shared structure. Van Fraassen makes this connection by using embeddability as a guarantee of the shared structure between theoretical models and "the phenomena", maintaining that "certain parts of the [theoretical] models [are] to be identified as empirical substructures, and these [are] the candidates for representation of the observable phenomena which science can confront within our experience" [82, p. 227]. This empiricist version of scientific structuralism avoids the question of why it should be assumed that the phenomenon is *represented* by data models by simply collapsing any distinction between the two and so offers no justification for why such an identification should be presumed possible".

[58] For example, for the ontic structural realist like French the appropriateness of the models of a theory is supposed to rely upon "how the world is"; the structure of the phenomena *qua* "nothing but structure" tells us that the world and the models (both data and theoretical models) share the same kind of structure. In contrast, for a structural empiricist like [82], the models of a theory (if they are embeddable) tell us not about the structure of "the world", but rather speak to the appropriateness of the theory for making claims about what we can know about the structure of the phenomena, i.e., it tells us about the empirical adequacy, but not the truth, of the theory.

[59] See Brading and Landry for the details and problems of each of these stances.

mark a significant cut-off point for the scientific structuralist; below the level of data models we seem to require more than analyses of shared structure between models, both to talk about the structure of the phenomena and to connect this to the structure of the world. In what follows, to avoid those philosophical problems associated with talk of "the phenomena", I place my focus on what can be said of the shared structure between data models and the world.[60]

Brading and Landry [8] separate the scientific structural realist's "philosophical" challenge of establishing a theory-world connection into two tasks: (i) to give an account of *applicability* in terms of the shared structure between models of the theory and data models wherein a theory, as a hierarchy of models, *presents what we say about* the structure of the world in terms of the *kinds of objects* that the data models are intended to *talk about*, and (ii) to give an account of *representation* in terms of the shared structure between the structure of the data models and the structure of the world wherein a theory, again as a hierarchy of models, *represents what we say* about the structure of the world in terms of the *particular objects* that the data models are intended to *be about*. To meet the requirements of the first task, we offered-up our *methodological*, or *minimal, scientific structuralism*, whereby one

> is committed only to the claim that the kinds of objects that a theory talks about are presented through the shared structure of its models and that the theory applies to the phenomena [as structured data] just in case the theoretical models and the data models share the same kind of structure. No ontological commitment—nothing about the nature, individuality, or modality of particular objects—is entailed. (p. 577)

In what follows, I plan to use NMA to meet the requirements of the second task and so offer-up a methodological structural *realist* position. But in contrast to its use by French, I will claim that this "argument"[61] only has a use at the "object level" and not at any "meta-level". This because, unless we presume that the world is set-structured, appeals to *the* meta-linguistic structure of scientific *theories* have no ontological bite. That is, there is no *global*, meta-level, "philosophical" perspective of the structure of scientific theories from where NMA can be put to use.[62] There is,

---

[60] This move, for example, allows us to side-step the problematic "philosophical" issues of how to understand what is meant by 'the phenomena' and where to place the phenomena in the Suppesian hierarchy [3, 54]. Van Fraassen too considers these issues, claiming: "the point long emphasized by Patrick Suppes that the theory is not confronted with the raw data but with models of the data, and the construction of these data models is a sophisticated and creative process" [82, p. 229]. However, he next leaves us philosophically confused by claiming that models of data are "the dress in which the debutante phenomena make their debut" [82, p. 229].

[61] I, like Worrall, (see Chapter 4) will presume that NMA is best taken as an "intuition", as opposed to as an argument, so that "the No Miracles intuition does no more, though also no less, than set some sort of realism as the default position and that it needs no more formal representation in order to do so".

[62] I take it that the object/meta-level distinction found in French (Chapter 1) thus matches onto the global/local use of NMA and so matches onto the retail/wholesale use of NMA found in Worrall (again, see Chapter 4).

however, a *local*, object-level, "scientific" perspective of the structure of a *particular* scientific theory from where NMA can be used to "cut nature".[63]

Before continuing on to suggest how one can be a methodological scientific structuralist *and a realist*, I again pause to summarize what's come before. Like philosophers taking the linguistic turn, my focus is on the content and structure of what we say, as opposed to the constitutive, "philosophy first", means by which we come to say them. Like the logical positivist, I eschew metaphysics as such a constitutive means and focus instead on the language of mathematics itself as the vehicle for the expression of what we say. Unlike the logical positivist, I eschew any need for a meta-linguistic framework, either logical (e.g., type-theoretic) or mathematical (e.g., set-theoretic). It is in this sense, then, that I take a scientific theory, as a hierarchy of physically interpreted, mathematically expressed, theories, as a Carnapian "linguistic framework".

Like the logical empiricist, I hold that the heart of the matter is empiricism, that is, the question of the extent to which *what we say* matches *what exists*, is answered by some "empirical" procedure or prediction. Yet, I eschew any epistemological, ontological, or linguistic distinctions that serve to underwrite our empirical "cutting of nature", e.g., phenomena/noumena, acquaintance/description, observables/unobservable terms or entities, detectable/auxiliary properties, etc. Instead I place my focus on what can be said of the match between what we say and what exists from *within* some scientific theory *qua* a hierarchy of structured systems. Thus, as a *methodological structuralist*, I take a scientific theory to *present* the content and structure of what we say about what exists, *and* this in virtue of the methods, rules and practices that are adopted in some context, and so as yielding the appropriate notion of shared structure that can be used "cut nature", so that, in the sense intended by logical empiricists, "structure yields objectivity".

Having situated my methodological scientific structuralism, I return to consider the case of the role of group theory in quantum mechanics, now with the aim of determining the proper place for the use of NMA as an "argument" for realism; more specifically, for the proper understanding of the claim that group-structure yields objectivity, and, indeed, "constitutes" objects.[64] French [30] discusses two "contexts

---

[63] That is, whereas Brading and Landry stopped at a methodological, or minimal, structuralist account of the *presentation* of *kinds of objects*, my aim here is to use a local version of NMA to extend this to a methodological structural realist account of the *representation* of *particular* objects.

[64] As Castellani [16, p. 14], notes, "[t]he use of the group-theoretical notion of invariance in relation to the object question is basically grounded on the idea that the possibility of speaking in terms of "objects" in a given context is connected with the possibility of individuating invariants with respect to the symmetry group of the context. This idea was first introduced by Felix Klein with regard to "geometrical objects", as a corollary of his new conception of geometry proposed in the famous 1872 *Erlanger Programm*. With the subsequent application of group theory to other domains of science and in particular to physics, this view could then be extended to other sort of "objects", and in particular to "physical objects" (the objects of physical theories as well as the objects of our common perception)." For example, in the context of theories of relativity, "objective is what is invariant with respect to the group of transformations of the spatio-temporal frame

of applicability"[65] of group theory in development of quantum mechanics. One, the 'Weyl programme', relates to the extent to which group theory can be used to *present* what we say about *the structure of quantum theory*. And, in so doing, if one is a *scientific structuralist*, one can appeal to group-structure as the "appropriate kind of structure" that can be used to "cut" the objects *of the theory* into *kinds of objects*, i.e., into kinds of objects *such that* they are appropriately group-structured. The other, the 'Wigner programme', relates to the extent to which quantum mechanics so presented can be further used to *present* what we say about *the structure of the data*. And, in so doing, if one is a *structuralist realist*, and appeals to NMA to argue that what we ought be realist about is, say, the group-structure of an object so "constructed", one can use this structure to "cut" the objects *of the world*. We can then say that the structure of the data *represents* the structure of the world, i.e., we may speak of a quantum mechanical object as a *particular* or *this such that* it is group-structured. The *ontic* structural realist (OSR) then claims that *all that there is* to the *this* is its *such that*, while the *epistemic* structural realist (ESR) claims that *all that is known* (or knowable) about the *this* is its *such that*. In any case, the claim of the structural realist is that we avoid the force of the pessimistic meta-induction argument (PMI), in so far as this argument rests on discontinuities, at the ontological level, of the *this*, by focusing instead on the continuity, at the structural level, of the *such that*.

Previously,[66] I argued that French's set-theoretic frame, and the appeal to partial structures and partial isomorphisms (or partial homomorphisms in the case of the shared structure of mathematical and physical theories), *does no real work*, either in *presenting* the structure of quantum mechanical *theories* or in *representing* the structure of quantum mechanical *objects*. I leave the interested reader to consider my argument. What I did not do, however, is explain the sense in which I do think that group-theoretic structure and group-theoretic morphisms, in fact, *do the real work*, and this work for *both* the *methodological structuralist's* account of *presentation* and for the *structural realist's* account of *representation*. As regards presentation, I did point out that

> there are two contexts that determine the appropriate kind of morphism and in each it is group-theoretic morphisms that do the work. The first context, of *reasoning from the phenomena*, is exemplified by Weyl's programme; it uses the 'relevant symmetries' to "work up" from the concrete structure of the phenomena [as structured data], via quantum mechanical principles and/or experimental results expressed as group-theoretic symmetries, to *present* the abstract structure of the theory. The second context, of *reasoning to the phenomena*, is exemplified by Wigner's programme; it uses the 'relevant symmetries' to

---

or, in the words of Hermann Weyl, "objectivity means invariance with respect to the group of automorphisms [of space-time]" [16, pp. 14–15].

65 This distinction between Weyl and Wigner is certainly over emphasized for the purpose of illustration, and so is somewhat unhistorical, but I use it to make my point against French. To more fully appreciate that Weyl was as concerned as Wigner with the use of group theory in the application of quantum mechanics to empirical data see [83].

66 See [47].

> "work down" from the abstract mathematical theory, via the group-theoretic formalism and corresponding 'internal bridges,' to *present* the concrete structure of the phenomena [as structured data]. In either case, what does the *real work* is the group-theoretic morphisms that underwrite, in both the foundational and representational contexts, the 'relevant symmetries,' and so serve to tell us what the appropriate kind of structure is for both presenting the structure of quantum theory and for representing the structure of the quantum phenomena [as structured data]. [45, p. 16]

Thus, in virtue of expressing quantum mechanics group-theoretically, this theory presents the structure of quantum mechanical objects, e.g., it presents bosons and fermions as group-theoretically "constituted" *kind of objects.* This, however, is where the methodological structuralist story ends. That is, we are left with a top-down or, as Brading put it, a law-constitutive account[67] of kinds of objects. As French explains,

> [w]e begin with a conceptualization of the phenomena informed by a broadly classical metaphysics in terms of which the entities involved are categorized as individuals. That categorization is projected into the quantum domain, where it breaks down and the fracture with the classical understanding is driven by the introduction of group theory; the entities are classified via the permutation group which imposes perhaps the most basic division into 'natural kinds', namely bosons and fermions. It is over this bridge that group theory is related to quantum mechanics as indicated above. [32, p. 204]

Again, French, as an ontic structural realist, uses this "top-down" approach, in combination with an appeal to a *global* version of NMA, i.e., to an NMA that ranges over the structure of scientific theories, to further argue that (set) structure "cuts nature", i.e., that *objects themselves* either are *nothing but* their structure or their structure is in some sense *ontologically prior.* I, as a methodological structuralist, used this, without any appeal to NMA, to make the more modest claim that for a particular theory of quantum mechanics group-theoretic structure is used to impose[68] *kinds of* objects.

Physically interpreted group theory then is our "linguistic framework"; it is used to *present* the content and structure of *what is said* both at the theoretical level and at the level of the data. I further made the distinction between presentation

---

[67] Note that Born [5] also made use of the notion of a law constitutive account of objects. Borrowing from this history, Castellani [16, p. 16], notes: "The important point is that the application of group-theoretical methods to contemporary physics has made it possible, in some way, to derive these invariant properties [mass, spin, charge] on the ground of symmetry considerations. In other words, we are provided with a general procedure for "constructing" or "constituting" the objects of physical theories as sets of invariants. This possibility supplies the basic motivation for what may be called the group-theoretical approach to the problem of defining physical objects, a recently developing area of inquiry centered on the invariance idea and its exploitation by using the results of the application of group theory to contemporary physical theories". For a well worked-out example of such "object constitution" within the group-theoretical approach, namely the group-theoretic construction of classical and quantum particles in the non-relativistic (i.e., "Galilean") case, see [15]. For a history of these ideas of see both [83] and [5].

[68] More to the point what imposes these kinds is the group symmetries (again, see [15]), and so we say that group theory is the language that best expresses these physical symmetries so that the physical system that presents these kinds shares structure with the data.

and representation by noting that from within our linguistic framework we have methodological principles bearing on two types of tasks (corresponding to the Weyl and Wigner programmes): one bearing on the *construction of the theory* and the other on the *description of the empirical data.* I noted too that the methodological principles and rules that are used to "carve out" the necessary properties and relations, and the laws so expressed in terms of these that act as the "bridge" between these two tasks, are both expressed group-theoretically[69] and given by the practices engendered by the work of scientists, not that of philosophers. But again, this only gets us to methodological structuralism.[70]

Now let's consider whether, and to what extent, we can use a *local* version of NMA—a version of NMA that only considers the extent to which a *particular* scientific theory presents the content and structure of what we say about what exists, and this in virtue of the methods, rules and practices that are adopted in some context, and so yielding the appropriate notion of shared structure that can be used "cut nature" (as opposed to just "imposing kinds") and so move us beyond methodological structuralism to structural realism. Recalling again our various stances; (i) from a *purely methodological stance*, we may forgo talk of *the* structure of the world and simply begin with structured data, i.e., with data models structured by the kinds that the theory presents; (ii) from an *empirical stance* we may say that what structures the phenomena into data models is the high level theory; and finally, (iii) from a *realist* stance we may say that what structures the phenomena into data models is the world. So far, as methodological structuralists, we have that the structure of the theory "cuts" *what we say* about the structure of the world by *presenting* us with *kinds of objects.* That structural *realist* question is: How do we know that this structure "cuts nature", i.e., "cuts" *what exists*, and so *represents* the structure of the world? How do we get from the *presentation of the structure of the theory or data*, i.e., from objects *such that* they are structured group-theoretically into the *kinds* 'boson', 'fermion', etc., to the *representation of the structure of the world*, i.e., to *this particular* object that *is* a boson or *is* a fermion.

As any realist should, we need to talk about objects; what can we know about them and what are they? I have said that our *expressions of* the existence of objects, in terms of the kinds of objects that a physical theory *presents*, are made from within a linguistic framework, so that, objects, as Castellani [15] and Brading [6] claim, are "law constitutive" or, as French explains, are imposed "top-down". I have also said that physical theories, insofar as they are the mathematical expressions of the

---

69 More specifically, the group-theoretic presentation of the action of symmetry transformation in Hilbert space yields the rules of transformation which itself yields the "quantum numbers" (for charge, spin, mass). It is these, moreover, that allow us to characterize what an electron is as a kind of object, yet, not what it is as an individual.

70 This point I take as in agreement with Psillos' [60] claim that Worrall's [85] use of the Fresnel light example, without properly situating his use of NMA and by relying on a problematic distinction between nature (content) and structure (form), is best read as providing "methodological insight", i.e., is best read as relating to the *presentation* of the structure of scientific theories, and not relating to the *representation* of the structure of the world.

content and structure of what we say, present objects as kinds and not as particulars; e.g., they present *what we say* about *any* or *all* bosons and fermions *such that*, but yet are not *about this* boson or *this* fermion *such that*. But too, for the structural realist, physical theories are not only presentations but are intended to be *representations*; they not only are used to *talk about* kinds of objects as positions in a structured physical system, they are intended to be *about* objects in the world. Thus, I note that while the argument for the presentation of objects as kinds that structure the data is "top-down", the argument for the representation of particular objects that are structured by the world needs be "bottom-up". That is, it needs to come via some claim of empirical/predictive success which "strengthens our conviction"[71] of a match or correspondence with the world.[72] Here, then, is where NMA must do its work.

OSRs have argued that there is no such gap between *talking about a kind of object* and *being about a particular object* because all there is to an object is its structure, i.e., *all there is* to a particular object as a *this* is its *such that*,[73] and since, by a *global* appeal to NMA, that is what is continuous over theory change, we're done! I, like the ESR, accept the gap[74] between a *this* and a *this such that*, claiming instead that *all we know* of a object as a *this* is as a kind of object *such that*, i.e., that it has such and such properties and relations in virtue of its being presented as a position in a physically interpreted mathematically structured system. But I note too that it is precisely here that a *local, contextual* version, of NMA does its work. That is, it would be a miracle if *a* mature, empirically/predictively successful theory, by presenting what we say about a kind of object as an object *such that*, did not match, correspond, or *represent*, in some sense, the world, i.e., did not represent a particular object as *this such that*.[75] And, for the methodological structural realist, who places his focus on "the appropriate kind of structure" that is both determined

---

71 See Duhem [21, p. 28], who says: "The highest test, therefore, of holding a classification as a natural one is to ask it to indicate in advance things which the future alone will reveal. And when the experiment is made and confirms the predictions obtained from our theory, we feel strengthen in our conviction that the relations established by our reason among abstract notions truly corresponds to relations among things".

72 Recall though that the structural realist is no more committed to truth, and so to a "correspondence theory of truth" than is the non-realist. That is, the notion of approximate truth will do just fine here as an underpinning for what we mean by 'match' or 'correspondence'.

73 This can be read as an ontological claim and thus we have what is standardly called *eliminative* OSR or as a claim of "reconceptualization", thus yielding *non-eliminative* OSR.

74 But, I take this gap to be a methodological consequence of the distinction between presentation and representation. Unlike for Worrall [85], it is neither an ontic nor an epistemic gap that is determined by, for example, the nature/structure distinction or the noumena/phenomena distinction.

75 One might worry how NMA can by itself get one from representation of a this such that to representation as a this such that *here and now*. For example, van Fraassen has been insisting that there also has to be some kind of indexical element in order to get to at a this. I deny that indexicality is a component of an object as a particular this, it might well be a component of *our perspective* of a this, but this is yet another matter altogether. Thanks to Paul Teller for raising this worry.

by the methods, practices, rules, principles and laws that scientists use and that is continuous over successively successful theories, the "some sense", is to be located in the shared structure between the kind and the particular, i.e., in the *such that*, not in the *this*.

Now it might be objected; how can you talk about a particular object as a *this* without "individuating criterion", that is, without countenancing particulars as, in some sense or other, individuals, entities, etc. Here I point the reader to Chapter 5, of this volume.

> For French and Ladyman, a realism that commits us to physical objects, but fails to determine the individuality or otherwise of those objects, is so strange that they reject it in favor of a commitment to "pure structure" as ontologically basic. Our view is that individuality is distinct from object-hood, and that the "metaphysical underdetermination worry" over individuality can be avoided in a less dramatic sounding manner. By appealing to the law-constitutive account of physical objects, we can pull apart objecthood and individuality in a very natural way: if a theory makes no commitments concerning whether or not the objects it purports to be about are individuals, then we need not conjoin a metaphysics of individuals versus nonindividuals to that theory in order to have a physical notion of object for our theory to be about. In such a case, requiring that we discuss these objects in terms of individuality (and perhaps even commit ourselves one way or another on the matter) demands that we go beyond the content of the theory: we have to add an interpretational layer not warranted by what the theory itself says. Expressed in this way, the alleged "strangeness" of a commitment to objects that is not accompanied by a metaphysics of individuality doesn't sound strange at all – at least not to us.

In either case, both Psillos' claim that structuralism (mathematical and scientific) requires metaphysical, or at least pre-theoretic, objects *qua* individuals, entities, etc., which would have us reify the *this* over the *such that*, and the ontic structural realist "argument from the metaphysical underdetermination",[76] which would have us eliminate or reconceptualise[77] *this* as only a *such that*, are both scuttled. This, then, is the sense in which, we may claim, without appeal to any meta-linguistic framework, or metaphysical ontology, that, from the level of the "language of science", structure yields both objectivity and, indeed, by appeal to a local NMA, yields objects.[78]

So NMA, when pointed at the "appropriate kind of structure", certainly can be used to pick out the structure that has "carving power", but these structuralist scissors only work when they are aimed locally, that is, when the 'intuition' behind NMA is directed at claims made from within the context of some scientific theory *qua* a "linguistic framework" and too when these claims are considered in light of the methodological principles, rules and laws, etc., that scientific practice estab-

[76] For example, in quantum mechanics, the permutation symmetries yield kinds of objects *qua indistinguishable particles* though *not* as individuals. See [15].

[77] Again, the eliminative view would have us eliminate objects entirely whereas the reconceptualized view would have us range over "objects" *qua* equivalence classes.

[78] To make the connection with this claim and my methodological reconstruction of Carnap, refer back to page 54. For more on what I intend by this in the case of quantum mechanical particles, see footnotes 62, 65 and 73.

lishes as needed for the bridge[79] between the construction of this framework for the structure of the theory and the description of the structure of the world. Thus, while the *presentation* of quantum theory and the shared group-theoretic structure of its models only gets us to bosons and fermions as *kinds of objects*, yet, in light of a local appeal to MNA, we can get to claims about the *representation* of the structure of the world, and so can be realists about the group-structure of bosons and fermions as particular objects in the world. This, then, is the sense in which I offer MSR as the best of the worlds between OSR and ESR.

References

1. Benacerraf, P. (1965) What Numbers Could Not Be. In P. Benacerraf and H. Putnam (eds.), *Philosophy of Mathematics*, 2nd ed. (pp. 272–294). Cambridge: Cambridge University Press, 1991.
2. Bernays, P. (1967) Hilbert, David. In P. Edwards (ed.), *The Encyclopedia of Philosophy,* Vol. 3 (pp. 496–504). New York: Macmillan.
3. Bogen, J. and J. Woodward (1988) Saving the Phenomena. *The Philosophical Review* **XCVII**: 303–352.
4. Bokulich, A. (2003) Horizontal Models: From Bakers to Cats. *Philosophy of Science* **70**: 609–627.
5. Born, M. (1953) Physical Reality. In E. Castellani (ed.), *Interpreting Bodies: Classical and Quantum Objects in Modern Physics* (pp. 187–203). Princeton: Princeton University Press, 1998.
6. Brading, K. (2011) Structuralist Approaches to Physics: Objects, Models and Modality. In A. Bokulich and P. Bokulich (eds.), *Scientific Structuralism* (pp. 43–65). Boston Studies in the Philosophy of Science, 281. New York: Springer.
7. ———— (2010) On Composite Systems: Newton, Descartes and the Law-Constitutive Approach. In P. Anstey and D. Jalobeanu (eds.), *Vanishing Matter and the Laws of Motion Descartes and Beyond* (pp. 130–152). New York: Routledge.
8. ———— and E. Landry (2007) Scientific Structuralism: Presentation and Representation. *Philosophy of Science* **73**(5): 571–581.
9. Carnap, R. (1966) The Nature of Theories. In M. Gardner (ed.), *Philosophical Foundations of Physics: An Introduction to the Philosophy of Science* (pp. 225–246). New York: Basic Books.
10. ———— (1956) Empiricism, Semantics, and Ontology. In P. Benacerraf and H. Putnam (eds.), *Philosophy of Mathematics* (pp. 241–257), 2nd ed. Cambridge: Cambridge University Press, 1991.
11. ———— (1934) *The Logical Syntax of Language*, A. Smeaton (trans.). London: Kegan Paul, 1937.
12. ———— (1928) *The Logical Structure of the World*, R.A. George, (trans.). London: Routledge & Kegan Paul, London, 1967.
13. Cartwright, N. (1999) *The Dappled World: A Study of the Boundaries of Science.* Cambridge: Cambridge University Press.
14. Castellani, E. (2002) Reductionism, Emergence, and Effective Field Theories. *Studies In History and Philosophy of Modern Physics* **33**: 251–267.

[79] Note that I using the term bridge here in the sense given in the French quote used on page 54, e.g., in the sense that the permutation group acts as a bridge that "imposes perhaps the most basic division into 'natural kinds', namely bosons and fermions". I am not, then, using this term in any way like what might use the term in sense of 'bridge principle' or 'correspondence rule'.

15. ———— (1998a) Galilean Particles: An Example of Constitution of Objects. In E. Castellani, (ed.), *Interpreting Bodies: Classical and Quantum Objects in Modern Physics.* Princeton: Princeton University Press.
16. ————, ed. (1998b) *Interpreting Bodies: Classical and Quantum Objects in Modern Physics.* Princeton: Princeton University Press.
17. Chakravartty, A. (2007) *A Metaphysics for Scientific Realism: Knowing the Unobservable.* Cambridge: Cambridge University Press.
18. ———— (2003) The Structuralist Conception of Objects. *Philosophy of Science* **70**: 867–878.
19. ———— (1998) Semirealism. *Studies in History and Philosophy of Science* **29**: 391–408.
20. Coffa, J. A. (1993) *The Semantic Tradition for Kant to Carnap: To the Vienna Station.* Cambridge: Cambridge University Press.
21. Duhem, P. (1906) *The Aim and Structure of Physical Theory*, P. Wiener (trans.). Princeton: Princeton University Press, 1954.
22. Ewald, W. (1996) *From Kant to Hilbert: A Source Book in the Foundations of Mathematics, Volume II.* Oxford: Oxford University Press.
23. Feigl, H. (1950) Existential Hypotheses: Realistic versus Phenomenalistic Interpretations. *Philosophy of Science* **17**: 35–62.
24. Frege, G. (1884) *The Foundations of Arithmetic*, J.L. Austin (trans.). Evanston, IL: Northwestern University Press, 1986.
25. ———— (1882) Boole's Logical Formula-language and my Concept-script. In P. Long and R. White (eds.), *Posthumous Writings* (pp. 47–52). Oxford: Blackwell, 1979.
26. ———— (1880/81) Boole's Logical Calculus and the Concept-script. In P. Long and R. White (eds.), *Posthumous Writings* (pp. 9–46). Oxford: Blackwell, 1979.
27. ———— (1879) Begriffsschrift. In T.W. Bynum (trans. & ed.), *Conceptual Notation and Related Articles* (pp. 101–203). Oxford: Oxford University Press, 1972.
28. Friedman, M. (1994) *Kant and the Exact Sciences,* 2nd ed. Cambridge, MA: Harvard University Press.
29. French, S. (2006) Structure as a Weapon of the Realist. *Proceedings of the Aristotelian Society* **106**: 1–19.
30. ———— (2000) The Reasonable Effectiveness of Mathematics: Partial Structures and the Application of Group Theory to Physics. *Synthèse* **125**: 103–120.
31. ———— (1999) Theories, Models and Structures: Thirty Years On. *Philosophy of Science, Supplement: Proceedings of PSA 1998.*
32. ———— (1999) Models and Mathematics in Physics: The Role of Group Theory. In J. Butterfield and C. Pagonis (eds.), *From Physics to Philosophy* (pp. 187–207). Cambridge: Cambridge University Press.
33. ———— and J. Ladyman (2003) Remodelling Structural Realism: Quantum Physics and the Metaphysics of Structure. *Synthèse* **136**(1): 31–56.
34. Glymour, B. (2000) Data and Phenomena: A Distinction Reconsidered. *Erkenntnis* **52**: 29–37.
35. Goldfarb, W. (2001) Frege's Conception of Logic. In J. Floyd and S. Shieh (eds.), *Futures Past: Reflections on the History and Nature of Analytic Philosophy* (pp. 25–43). Cambridge, MA: Harvard University Press.
36. Hallett, M. (2007) Reflections on the Purity of Method in Hilbert's. *Grundlagen der Geometrie*. Unpublished manuscript.
37. ———— (1994) Hilbert's Axiomatic Method and the Laws of Thought. In A. George (ed.), *Mathematics and Mind* (pp. 158–200). Oxford: Oxford University Press.
38. Hawthorne, J. (2001) Causal Structuralism. *Philosophical Perspectives* **14**: 361–78.
39. Hellman, G. (2001) Three Varieties of Mathematical Structuralism. *Philosophia Mathematica* **3**(9): 184–211.
40. ———— (1996) Structuralism Without Structures. *Philosophia Mathematica* **2**(4): 100–123.
41. Kneale W. and M. Kneale (1984) *The Development of Logic.* Oxford: Oxford University Press.

42. Ladyman, J. (2007) On the Identity and Diversity of Objects in a Structure. *Proceedings of the Aristotelian Society* **81**: 23–43.
43. ———— (2002) *Understanding Philosophy of Science*. New York: Routledge.
44. ———— (1998) What is Structural Realism? *Studies in the History and Philosophy of Science* **29**: 409–424.
45. Landry, E. (2010) A Silly Answer to a Psillos Question. In D. DeVidi, M. Hallett, and P. Clarke (eds.), *Logic, Mathematics, Philosophy, Vintage Enthusiasms: Essays in Honour of John Bell* (pp. 361–389). Western Ontario Series for Philosophy of Science. New York: Reidel.
46. ———— (2009) How to be a Structuralist All The Way Down. *Synthèse* **179**: 435–454.
47. ———— (2007) Shared Structure Need Not Be Shared Set-Structure. *Synthèse* **158**: 1–17.
48. ———— (2006a) Category Theory as a Framework for an *In Re* Interpretation of Mathematical Structuralism. In J. van Benthem, G. Heinzmann, M. Rebuschi, and H. Visser (eds.), *The Age of Alternative Logics: Assessing Philosophy of Logic and Mathematics Today* (pp. 163–177). New York: Kluwer.
49. ———— (2006b) Intuition, Objectivity and Structure. In E. Carson and R. Huber (eds.), *Intuition and the Axiomatic Method* (pp. 133–153). Western Ontario Series for Philosophy of Science. New York: Reidel.
50. ———— (1999) Category Theory: The Language of Mathematics. *Philosophy of Science* **66** Supplement: S14–S27.
51. ———— and J.-P. Marquis (2005) Categories in Context: Historical, Foundational and Philosophical. *Philosophia Mathematica* **13**(1): 1–43.
52. Maddy, P. (2007) *Second Philosophy*. New York: Oxford University Press.
53. Mancosu, P., ed., (1998) *From Hilbert to Brouwer. The Debate on the Foundations of Mathematics in the 1920s*. Oxford: Oxford University Press.
54. McAllister, J. (1997) Phenomena and Patterns in Data Sets. *Erkenntnis* **47**: 217–228.
55. Morgan, M. and M. Morrison, eds. (1999) *Models as Mediators: Perspectives on Natural and Social Science*. Cambridge: Cambridge University Press.
56. Parsons, C. (1990) The Structuralist View of Mathematical Objects. *Synthèse* **84**: 303–346.
57. Psillos, S. (2006) The Structure, the Whole Structure and Nothing but the Structure? *Philosophy of Science* **73**(5): 560–570.
58. ———— (2001) Is Structural Realism Possible? *Philosophy of Science* **68**: 13–24.
59. ———— (1999) *Scientific Realism: How Science Tracks Truth*. London: Routledge.
60. ———— (1995) Is Structural Realism the Best of Both Worlds? *Dialectica* **49**: 15–46.
61. Quine, W.V.O. (1951) Two Dogmas of Empiricism. In W. D. Hart (ed.), *The Philosophy of Mathematics* (pp. 31–51). Oxford: Oxford University Press, 1996.
62. Richardson, A. (1998) *Carnap's Construction of the World*. Cambridge: Cambridge University Press.
63. Rickles, D. and S. French (2006) Quantum Gravity Meets Structuralism: Interweaving Relations in the Foundations of Physics. In S. French, D. Rickles, and J. Saatsi (eds.), *The Structural Foundations of Quantum Gravity* (pp. 1–39). Oxford: Oxford University Press.
64. Russell, B. (1948) *Human Knowledge: Its Scope and Limits*. London: George Allen and Unwin, 2009.
65. ———— (1918) The Philosophy of Logical Atomism. In *Logic and Knowledge* (pp. 177–283). New York: Capricorn Books.
66. Saatsi, J. (2005) Reconsidering the Fresnel-Maxwell Theory Shift: How the Realist can have her Cake and EAT it too. *Studies In History and Philosophy of Science* **36**: 509–538.
67. Saunders, S. (2006) Are Quantum Particles Objects? *Analysis* **66**: 52–63.
68. ———— (2003) Physics and Leibniz's Principles. In K. Brading and E.Castellani (eds.), *Symmetries in Physics: Philosophical Reflections* (pp. 289–308). Cambridge: Cambridge University Press.
69. ———— (2003) Structural Realism, Again. *Synthèse* **136**: 127–133.
70. Schindler, S. (2007) Rehabilitating Theory: Refusal of the 'Bottom-up' Construction of Scientific Phenomena. *Studies in History and Philosophy of Science* **38**: 160–184.

71. Shapiro, S. (2005) Categories, Structures, and the Frege-Hilbert Controversy: The status of Meta-Mathematics. *Philosophia Mathematica* **13**(1): 61–77.
72. ———— (1997) *Philosophy of Mathematics: Structure and Ontology.* Oxford: Oxford University Press.
73. ———— (1996) Space, Number and Structure: A Tale of Two Debates, *Philosophia Mathematica* **3**(4): 148–173.
74. Suppes, P. (1967) *Set-Theoretical Structures in Science.* Stanford: Stanford University Press.
75. Suppes, P. (1962) Models of Data. In E. Nagel, P. Suppes, and A. Tarski (eds.), *Logic, Methodology and Philosophy of Science* (pp. 253–261). Stanford: Stanford University Press.
76. ———— (1960) A Comparison of the Meaning and Uses of Models in Mathematics and the Empirical Sciences. *Synthèse* **12**: 287–301.
77. Teller, P. (2004) How we Dapple the World. *Philosophy of Science* **71**: 425–447.
78. ———— (1998) Quantum Mechanics and Haecceities. In E. Castellani (ed.), *Interpreting Bodies: Classical and Quantum Objects in Modern Physics* (pp. 114–141). Princeton: Princeton University Press.
79. ———— (1995) *An Interpretative Introduction to Quantum Field Theory.* Princeton: Princeton University Press.
80. Toretti, R. (1999) *The Philosophy of Physics.* Cambridge: Cambridge University Press.
81. van Fraassen, B. (2002) *The Empirical Stance.* New Haven, Yale University Press.
82. ———— (1989) *Laws and Symmetry.* Oxford: Oxford University Press.
83. Weyl, H. (1952) *Symmetry.* Princeton: Princeton University Press.
84. Wittgenstein, L. (1921) *Tractatus Logico-Philosophicus,* D.F. Pears, and B.F. McGuinness (trans.). London: Routledge & Kegan Paul, 1992.
85. Worrall, J. (1989) Structural Realism: The Best of Both Worlds? *Dialectica* **43**: 99–124.

[illegible]

Chapter 3
How *Not* to Be a Realist

Ioannis Votsis

3.1 Introduction

When it comes to name-calling, structural realists have heard pretty much all of it. Among the many insults, they have been called 'empiricist anti-realists' but also 'traditional scientific realists'. Obviously the collapse accusations that motivate these two insults cannot both be true at the same time. The aim of this paper is to defend the epistemic variety of structural realism against the accusation of collapse to traditional scientific realism. In so doing, I turn the tables on traditional scientific realists by presenting them with a dilemma. They can either opt for a construal of their view that permits epistemic access to non-structural features of unobservables but then face the daunting task of substantiating a claim that up till now has failed to deliver the goods or they can drop the problematic requirement of epistemic access to non-structural features but then face a collapse to epistemic structural realism. There are good reasons to suspect that traditional scientific realists have, perhaps unwittingly, been edging towards the second option as some of their proclamations can attest. It is high time to let these epistemic structural realists out of the closet.

3.2 Epistemic Structural Realism

Structural realism is a factious family of related views in the scientific realism debate.[1] There are broadly speaking three kinds of structural realism: methodological, epistemic and ontic. Let us start with the methodological kind. This focuses on the role shared structure plays in characterising scientific theories, in relating high-level theory to low-level data and in identifying links between predecessor

[1] For a detailed critical survey see [8].

I. Votsis (✉)
Philosophisches Institut, University of Düsseldorf, Universitätsstraße 1,
40225 Düsseldorf, Germany
e-mail: votsis@phil.uni-duesseldorf.de

E.M. Landry, D.P. Rickles (eds.), *Structural Realism*, The Western Ontario Series in Philosophy of Science 77, DOI 10.1007/978-94-007-2579-9_3,

and successor theories (see Brading and Landry [3]). Although this is certainly a structural view, it is difficult to discern how this view earns the moniker 'realism'. No realist claim on the semantic, epistemic or ontic level is made by those who endorse this view and for this reason it would be better to classify it as a kind of structuralism (about the methodology of science) rather than as a kind of structural realism.

Consider next the ontic kind of structural realism or OSR for short. Several distinct versions of it exist. What all ontic structural realists have in common is the rejection of one or more claims associated with traditional conceptions of objects. In its original formulation (e.g. Ladyman [11]), what I call the 'no objects view', the position does away with objects and attempts to make do only with structures. That is, it calls for a reconceptualisation of ontology that sees objects merely as placeholders in structures. Another version of OSR that is perhaps the most prominent one is what I call the 'no individuals view' [7]. This view maintains the existence of objects but rejects that these should be conceived as individuals. Even though objects survive under this view what carries the ontological weight is once again the relations or structures.

Supporters of the epistemic kind of structural realism or ESR for short hold that although our knowledge of observables is unrestricted, our knowledge of unobservables is at best structural. In more formal terms, we can only know the unobservable world up-to-isomorphism. This view can be contrasted with traditional scientific realism whose advocates insist that both observable and unobservable aspects of the world are in principle fully knowable. In other words, the relevant difference between the two views is the extent to which unobservable aspects of the world can be known. There are two versions of ESR currently being sponsored. Those who endorse the Ramsey version claim that the structure of the unobservable world is best captured in the Ramsey sentence formulations of successful scientific theories [33].[2] Those who endorse the Russell version claim that we can infer certain things about the structure of the unobservable world from the structure of our perceptions [21, 27].

For the remainder of this paper my claims are solely concerned with ESR. Unless otherwise noted, my remarks will be largely blind to the two versions of ESR, i.e. they will apply equally to both of them. Having said this, it is worth mulling over the ways in which the two versions of ESR differ. Their differences can be plotted along three axes. First, there is the direct vs. indirect realism axis. Those advocating the Ramsey-sentence approach to ESR tacitly endorse direct realism. In so doing they claim that epistemic agents perceive, cognise and are aware of the world directly. For supporters of this view some but not all physical objects are unobservable. Subatomic particles are the clearest case of a class of unobservables. In opposition to this view, those advocating the Russellian approach to ESR endorse an indirect form of realism. They presuppose that the immediate object of our perception, cognition

[2] Many authors neglect the fact that in his original presentation of ESR Worrall [30] does not advocate the Ramsey sentence approach—indeed he makes no mention of it.

and awareness is something internal, e.g. a mental representation or at least some kind of by-product of the human perceptual system.[3] For supporters of this view the whole external world counts as unobservable and for that reason can at best be known structurally. In principle nothing stands in the way of marrying Ramsey-style ESR to indirect realism. The same cannot be said of Russell-style ESR for indirect realism is hardwired into it.

Both approaches to ESR are at variance with scientific realism on the issue of observability. Scientific realists deny that a clear line can be drawn between what is observable and what is unobservable or at least divest such a distinction of any epistemic significance. They do so because they want to undercut the empiricist anti-realist's attempts to motivate a selective agnosticism with respect to unobservables. ESR-ists also attempt to motivate a form of selective agnosticism, one that directs the agnostic attitude towards the non-structural features of unobservables. For this reason, ESR-ists are confronted with some of the same objections facing anti-realist empiricists. To be exact, only the direct realist approach to ESR is affected by such objections. This is because the objections question whether external world objects can legitimately be divided into separate (i.e. observable and unobservable) classes, a division the indirect realist rejects. Of course indirect realists have objections of their own to worry about. Alas, this is a discussion that needs to be put on ice for another occasion.

The second axis discriminates Ramsey-style from Russell-style ESR on the basis of how each view arrives at the much vaunted structure. Advocates of the former do so by translating successful scientific theories into their Ramsified counterparts where theoretical terms become existentially quantified variables. *Qua* variables they offer only structural clues about the individual objects that instantiate them. Psillos has called this the 'downward path to structural realism' in view of the fact that one starts with fully-fledged theories and then proceeds to peel away the non-structural elements, i.e. the intensions of the theoretical terms, to get to the structure. Compare this to what Psillos calls the 'upward path to structural realism', according to which we infer various things about the structure of the unobservables from the structure of the observables on the supposition that large parts of the two domains are isomorphic. This is the route taken by advocates of Russell-style ESR.

Finally, the third axis runs along the kind of arguments that have been utilised to motivate each view. Ramsey-style ESR has been motivated by arguments from the history of science. By contrast Russell-style ESR has been motivated by arguments from perception. This being said, nothing prevents one from mounting arguments from the history of science to support Russell-style ESR. Likewise nothing prevents one from mounting arguments from perception to prop up Ramsey-style ESR. What is more, some of these arguments can be cited to support other structuralist views. Arguments from the history of science, for example, were adopted and adapted early on by ontic structural realists.

[3] The identification of indirect realism with representationalism should be resisted. The latter is simply one manifestation of the former.

Before we turn to the subject at hand a formal account of the notion of structure is vital. Although the following set-theoretical account is not universally accepted, it is sufficiently widespread and gives enough of an intuitive grasp of what epistemic structural realists have in mind. A structure S is specified by two things: (i) a non-empty set U of objects, which is also known as the domain of the structure, and (ii) a non-empty indexed set R of (monadic and/or polyadic) relations defined on U. A structure so specified is a so-called 'concrete structure'. To understand the notion of structure that structural realists entertain we must abstract from this a notion of 'abstract structure'. This latter notion presupposes the idea of an isomorphic relation between structures. A structure $S_1 = (U_1, R_1)$ is *isomorphic* to a structure $S_2 = (U_2, R_2)$ just in case there exists a bijective mapping $f : U_1 \rightarrow U_2$ that preserves the system of relations between the two structures, i.e. for all relations $p_i \in R_1$ and $q_i \in R_2$, a set of objects $\{a_1, \ldots, a_n\}$ in U_1 satisfies the relation p_i if and only if the corresponding set of objects $\{b_1 = f(a_1), \ldots, b_n = f(a_n)\}$ in U_2 satisfies the corresponding relation q_i—the corresponding relations have the same index. We can now define the requisite notion: An *abstract structure* Σ is what all concrete structures that are isomorphic to one another have in common. Henceforth, and unless otherwise noted, talk of structure will denote talk of abstract structure.

3.3 Accusations of Collapse

An effective way to brush aside a viewpoint is to cast doubt on the distinctiveness of its character. Not only does this threaten to rob the given viewpoint of its originality but it also threatens to unload at its feet all the difficulties borne by the viewpoint it collapses into. In the case of ESR, two collapse accusations have been propounded. The first is precipitated by the notorious Newman problem. Named after its originator, the mathematician M.H.A. Newman, the problem zeros in on the way epistemic structural realists articulate their knowledge claims. To say, like they presumably do, that for a given class of unobservables there exists a system of relations with a certain logico-mathematical structure without at the same time identifying the specific relations is, according to Newman, to say nothing much since that same claim can be derived from theorems of set theory or second-order logic. The only supposition required for that derivation concerns the minimum number of objects in the given class. In other words, the only claim about the unobservable world left open for empirical determination, says Newman, concerns this cardinality supposition. Those who endorse the Newman problem take ESR as collapsing into a form of empiricist anti-realism for the only substantive knowledge claims it seems to make concern the observable world. If one is harbouring hopes of a robust form of realism, securing knowledge about the minimum number of unobservable objects can hardly be adequate. The Newman problem will not concern us further here—it has been widely discussed elsewhere [10, 31]—though I will come back to it briefly in Section 3.4 below.

The other major collapse accusation has hardly received any attention in the literature. In a nutshell, it is the accusation that epistemic structural realism places no realisable restriction on what can be known and hence collapses into traditional scientific realism. I here quote from Psillos, the prime mover of this accusation:

> ... to say what an entity *is* is to show *how this entity is structured*: what are its properties, in what relations its stands to other objects, etc. An exhaustive specification of this set of properties and relations leaves nothing left out. Any talk of something else remaining uncaptured when this specification is made is, I think, obscure [16, p. 156] [emphasis in original].

In this and adjoining passages Psillos grumbles about the epistemic structural realists' adherence to the existence of something which remains structurally unspecifiable and which they call the 'nature' of an entity. This use of the term 'nature' is in his eyes anachronistic.

> I think that talk of 'nature' over and above this structural description (physical and mathematical) of a causal agent is to hark back to medieval discourse of 'forms' and 'substances'. Such talk has been overthrown by the scientific revolution of the seventeenth century [16, pp. 155–156].[4]

Not having the same gripes with the notion of nature and the associated structure vs. nature distinction but sharing Psillos' intuition that epistemic structural realism, when properly construed, collapses to traditional realism, Papineau says:

> ... since our intellectual access to unobservable entities is always mediated by a structure of theoretical assumptions rather than by direct insight into their nature, Worrall's restriction of belief to structural claims is in fact no restriction at all [15, p. 12].

All in all, Psillos and Papineau agree that ESR's collapse to traditional scientific realism is effected by the former's inability to place a realisable restriction on what can be known.

ESR cannot collapse to both realism and anti-realism unless of course they are one and the same position, a supposition we are not entertaining here.[5] Oddly this tension seems to have remained undetected by Psillos and Papineau who endorse both collapse accusations. A scrupulous reader might at this point protest that the tension is only feigned since the Newman problem does not strictly speaking threaten to expose ESR as an anti-realist view but instead as an insufficiently realist one—recall that the Newman problem diagnoses ESR with a severely limited ability to assert anything non-trivial about the unobservable world. Be that as it may, the tension does not vanish but reappears on a different level. ESR cannot collapse to both an insufficiently realist view and a sufficiently realist one.

The new tension can be dissolved by expressing the two collapse claims as distinct options in a dilemma. This approach in fact follows the tenor of Newman's original critique. Either epistemic structural realists advocate a pure version of their

---

[4] I have dealt with Psillos' objections to the structure vs. nature distinction in my [28].

[5] It is not a-priori impossible that realism and anti-realism are ultimately identical positions. Such a suggestion is implicit in the work of some philosophers who wish to dissolve the scientific realism debate. Although there may be something to this suggestion, my target audience for this paper is those for whom the legitimacy of the scientific realism debate is not at issue.

view which collapses to some exceedingly weak form of realism or they advocate an adulterated version which collapses to traditional scientific realism. Those wondering what a pure version of ESR looks like need not look any further than the formulations of ESR given above. As for impure versions of ESR, let's just say for now that they are versions that profess knowledge of unobservables that goes beyond their structural features.

Whether pure ESR collapses to empiricist anti-realism or at least to some exceedingly weak form of realism is not a matter to be trifled with. In my view Russell's version of ESR is immune to the Newman objection. I have argued as much elsewhere [25, 26, chapter 4]. Let us suppose for the sake of the argument, however, that pure versions of ESR do indeed suffer this ignominious collapse. If this were true, epistemic structural realists would need to endorse an impure version of ESR. Would the mere shift to an adulterated version rob them of the originality of their view? Let us find out!

3.4 Adulterated ESR

Is the ESR dictum 'All we can know is structure' merely a catchy slogan that leaves out important qualifications? If so, do these qualifications conceal impurities that render ESR indistinguishable from traditional scientific realism? I have alluded elsewhere [26, p. 113] that an impure form of ESR need not be a capitulation to traditional scientific realism. Here I want to take a more sustained look at this issue.

Those who fancy the Newman problem as a knockdown argument against ESR (in any of its forms) often cite Russell's letter to Newman where he seems to sheepishly admit defeat:

> You make it entirely obvious that my statements to the effect that nothing is known about the physical world except its structure are either false or trivial, and I am somewhat ashamed at not having noticed the point for myself [23, p. 413].

It is utterly reprehensible, however, that these same people ignore what Russell goes on to say in that letter:

> It was quite clear to me, as I read your article, that I had not really intended to say what in fact I did say, that *nothing* is known about the physical world except its structure. I had always assumed spatio-temporal continuity with the world of percepts, that is to say, I had assumed that there might be co-punctuality between percepts and non-percepts, and even that one could pass by a finite number of steps from one event to another compresent with it, from one end of the universe to the other. And co-punctuality I regarded as a relation which might exist among percepts and is itself perceptible.
>
> I have not yet had time to think out how far the admission of co-punctuality alone in addition to structure would protect me from your criticisms, nor yet how far it would weaken the plausibility of my metaphysic. What I did realise was that spatio-temporal continuity of percepts and non-percepts was so axiomatic in my thought that I failed to notice that my statements appeared to deny it. [23, p. 413] [emphasis in original].

Russell reminds Newman that additional elements are required to make ESR stick and points out one of them—the assumption that percepts are spatiotemporally continuous with their causes.[6] This assumption is not opportunistically dreamt up by Russell but plays an integral role in his philosophy (see [21], chapter 21). Although an interesting matter in its own right, we will not here judge the assumption's warrantability or indeed its presumed indispensability for ESR. We simply note that Russell took it to be a central feature of ESR that, by his own admission, seems to introduce certain impurities into the position.

Before we scrutinise this thought, I want to momentarily direct the reader's attention to another erroneous belief propagated in the ESR literature. Consider the following remark from the *Routledge Encyclopedia of Philosophy's* entry on Russell:

> Russell quickly abandoned [E]SR when Newman showed that any set with the right cardinality could be arranged so as to have the same structure as the world—a result analogous to that claimed in Putnam's model-theoretic argument against realist theories of reference (Demopoulos and Friedman 1989) [4, p. 400].

Nothing could be further from the truth. Russell continued to highlight the structural nature of knowledge in much of his subsequent work. Take, for example, the following passage from *Human Knowledge*, published 20 years after Russell's letter to Newman:

> Anticipating coming discussions, I shall assume that the physical world, as it is independently of perception, can be known to have a certain structural similarity to the world of our percepts, but cannot be known to have any qualitative similarity [22, p. 138].

The above is one of many passages that demonstrate Russell's continued loyalty to ESR. Several Russellian scholars confirm this view, documenting his reliance on structuralist ideas long after the letter was sent to Newman—one good source is [2].

Let us now return to the question whether the spatiotemporality assumption introduces impurities into ESR, regardless of Russell's own thoughts on the matter. For something to count as an impurity in the current context it must add to the position's epistemic commitments, i.e. to the claims one is willing to endorse as knowledge. Does the spatiotemporality assumption do that? The answer to this question is rather unclear. The assumption is metaphysical in character, for it tells us something about the kind of world we are living in. The question then is whether our endorsement of it somehow rationally compels us to include it in our epistemic commitments. On the one hand, it may be argued that some metaphysical assumptions are required to get any epistemological project off the ground, even though we *do not* and perhaps *cannot know* that the world satisfies them. According to this approach, the spatiotemporality assumption is needed to secure a correspondence between the world we perceive and the world we live in but it cannot strictly speaking be included in our list of epistemic commitments[7]. The upshot of all this is that ESR remains unadulterated. On the other hand, it may be argued that our metaphysical commitments

---

[6] Russell in fact advocated a more general version of this principle, namely that all events are spatiotemporally continuous. The special case of the principle is established once one takes into account that percepts as well as their unobservables causes are events in his view.

[7] To maintain some measure of perceptual veridicality even those who reject ESR must accept some such assumption.

should never exceed our epistemological ones. Why, after all, should the warrant required for a given claim to become part of our metaphysical commitments be any different from the warrant required for a given claim to become part of our epistemic commitments? The upshot in this case is that ESR becomes adulterated.

Suppose for the sake of the argument that the spatiotemporality assumption introduces impurities into ESR. Does this automatically mean the collapse of ESR to traditional scientific realism? This question is easier to answer and the answer is 'No'. For ESR to collapse to traditional scientific realism, the inclusion of the spatiotemporality assumption into our list of epistemic commitments would have to bring with it the ability to fully specify the contents of one or more unobservable domains. I fail to see how the said assumption can achieve this feat. At best, the spatiotemporality assumption provides a very *general* constraint that all unobservable domains must obey. The same point applies to the 'impurities' cited by Psillos [18]. According to him, physical objects possess some knowable non-structural properties, namely 'that they are not abstract entities, that they are in space and time, that they have causal powers' (p. 567). Even if these properties are indeed non-structural and hence additives to pure ESR, I fail to see how they can bring about the full specification of the contents of one or more unobservable domains. This is because the aforesaid properties are presumably possessed by *all* physical objects. They are *not specific* to individual objects and therefore they cannot grant such objects their unique character. In sum, the kind of impurities ESR may be forced to endorse is not the kind that supports a collapse to traditional scientific realism.

Is this conclusion limited only to those advocating the Russellian version of ESR? In other words, can the qualifications made by Ramsey-style epistemic structural realists be interpreted as introducing impurities and, if so, do these impurities force a collapse to traditional scientific realism? In Worrall's view [32] the Ramsey-sentence of a successful scientific theory expresses much more about the unobservable world than assertions about its cardinality. Among the entailments of a Ramsey-sentence, he argues, are several theoretical assertions that no anti-realist would be willing to endorse. How is this possible one may ask, if theoretical predicates are turned into existentially quantified variables? The answer, according to Worrall, is that not all assertions made with a purely observational vocabulary are observational in character. The mark of a real theoretical assertion, he contends, is our inability to directly check its truth value by observation. Since some assertions formulated in a purely observational vocabulary cannot be checked in this way they are, for all intents and purposes, theoretical.[8] On the supposition that Worrall is right, there is more distance between Ramsey-style ESR and empiricist anti-realism than previously thought. Moreover, it seems that this distance is not the result of shedding Ramsey-style ESR's pure form, for no genuine expansion of epistemic commitments has occurred. Worrall's analysis has instead prompted us to take a closer look at what the Ramsey-sentence of a theory entailed all along.

---

[8] One of his examples is the assertion 'Nothing is older than 6000 years old' in the theoretical dispute between the Darwinists and the Creationists.

Suppose for the sake of the argument that Worrall's elaboration of the Ramsey-sentence approach introduces impurities into ESR. Does the resulting form of realism collapse to traditional scientific realism? The answer once more seems to be 'No'. The traditional scientific realist underwrites not just the Ramsey-sentence of a successful theory, which is of course entailed by the unRamsified theory itself, but also the interpretations of the unRamsified theory's theoretical terms. The latter is something the Ramsey-style epistemic structural realist vehemently denies we have epistemic access to. For someone like Worrall interpreted theoretical terms are in effect specific non-structural components. Ramsey-style ESR cannot thus be accused of collapse to traditional scientific realism.

The message of this section is, I hope, plain and clear. Even versions of ESR adulterated with additional epistemic commitments do not suffer a collapse to traditional scientific realism.[9]

3.5 Specific Non-structural Theoretical Components

It is now time to consider in some detail the additional epistemic commitments scientific realists sanction. One piece of information that I hope surfaced in the course of the preceding section is that there are two kinds of epistemic commitments that adulterate ESR. The first kind consists of epistemic commitments that on their own do not seem to push ESR over the edge and into the territory of traditional scientific realism, e.g. the spatiotemporality assumption. The second kind consists of those epistemic commitments that are sufficient to support ESR's collapse to traditional scientific realism. We called the latter kind 'specific non-structural components'. This section explores the prospects of finding specific non-structural components we should be realists about, a prospect that if realised would naturally mean the end of ESR.

Some scientific realists explicitly aver epistemic access to specific non-structural components of unobservables. Psillos [16, chapter 7], for example, asserts that specific theoretical components that are non-structural systematically survive theory-change. If correct, this assertion could potentially deal a devastating blow to ESR, for it would lend credence to the view that their survival is perhaps due to the essential role they play in the predictive and explanatory success of their respective theories—success being the ultimate sign for a theory's approximate truth or at least some kind of proximity to truth.[10] To properly evaluate Psillos' claim we need to comb through the history of science to ascertain: (i) whether specific non-structural theoretical components survive theory change and, if so, (ii) whether their survival

---

[9] This claim holds at least in so far as scientific realists explicitly endorse specific non-structural knowledge. Those scientific realists who do not endorse this claim are dealt with in Section 3.6.

[10] Theoretical components may of course survive theory change without playing an essential role in the predictive and explanatory success of their respective theories. Having said this, one expects to find a high degree of correlation between the survival of theoretical components and their integral role in the success of the theories they belong to for the simple reason that scientists generally aim to increase empirical success and eliminate idle wheels. For more on this see [29].

discloses a latching onto the world or is merely an accidental, convenient or conservative feature of the process of constructing a successor theory. Needless to say that question (ii) can be posed about any type of component survival through theory change, including that of structural components.

Psillos does not corroborate his claim with a systematic analysis of the history of science—a tall order for anyone. Instead he focuses on the case that made ESR famous, i.e. the transition from Fresnel's theory of light to Maxwell's theory of electromagnetism. To be exact, he focuses on a handful of assumptions that Fresnel apparently used to derive his laws of optics:

(a) A *minimal mechanical assumption* that the velocity of the displacement of the molecules of ether is proportional to the amplitude of the light-wave...
(b) The *principle of conservation of energy* ('*forces vives*') during the propagation of light in the two media...
(c) A *geometrical analysis* of the configuration of the light-rays in the interface of two media... [16, p. 158] [emphasis in original].

In Psillos' view, these three assumptions are 'fundamentally correct' for they purportedly survived theory-change, finding their way into Maxwell's electromagnetic theory. Moreover, they cannot be completely accounted for in structural terms. For this reason they provide some prima facie evidence in favour of traditional scientific realism as opposed to ESR.

Let us consider each of these assumptions in turn. The first one, the minimal mechanical assumption, states a mathematical relation between two quantities, viz. the amplitude of the wave and the velocity of the displacement of the ether molecules. Although this mathematical relation survives into the mature version of Maxwell's electromagnetic theory, its ontological import gets reinterpreted with the displacement of the ether molecules becoming a 'displacement' of the electromagnetic field strengths. Hence no specific non-structural theoretical component survives in this case.[11] What is even more puzzling about Psillos' appeal to (a) is that he eventually acknowledges that it is not really performing a substantive role in the derivation of Fresnel's laws. He thus says that the only assumption required in that derivation is to 'take energy as a function of the square of the amplitude of the light waves' [16, p. 159]. Indeed, Psillos reveals that Fresnel himself had recognised that 'no specific assumptions about the trajectories of the ethereal molecules were necessary' [16, p. 159].[12]

[11] In my view the wave's amplitude is not a theoretical component because it is the kind of quantity that can be measured, i.e. it is a broadly construed observable quantity. Its survival is thus no threat to ESR.

[12] Jonathan Bain also makes this point when he says that what Psillos calls the 'minimal mechanical assumption' 'was used solely to express the energy associated with a light-wave as the square of its amplitude with no essential reference to the medium of oscillation. Hence, again, one can argue that the aether was not used in the derivation' [1, p. 163].

The second assumption lends itself to a similar analysis. Jean Le Rond d'Alembert's account of the *forces vives*—or *vis viva* as it was better known—principle gives us an idea of what scientists at the time had in mind.

> If bodies act one against the other, either by pulling on threads or inelastic rods, by pushing or by impact, as long as in this last case it has perfect elasticity, the sum of the product of the masses multiplied by the square of the speeds will always be a constant quantity [6].

In other words, the principle asserts that the following quantity is conserved:

$$\Sigma_i m_i v_i^2 \tag{3.1}$$

where m_i indicate the masses of the bodies and v_i their corresponding velocities. Since the principle states a mathematical relation between masses and velocities, two measurable and hence broadly construed observable quantities, its survival through theory change leaves the epistemic structural realist unperturbed. Today we think of the *forces vives* principle as an attempt to formulate the idea that kinetic energy is conserved under elastic collisions.[13] We also have a more general principle of energy conservation, namely the conservation of total energy, which applies to both kinetic and potential energy.

The third and final assumption can also be dismissed rather easily. No realist supports the view that geometrical analysis represents any aspect of the world. Geometrical analysis is simply a tool that facilitates modelling and calculation. Its survival through scientific revolutions, therefore, has no epistemic significance for the realist, structural or other. Even if it had epistemic significance, I do not see how this would help Psillos' case since geometrical analysis involves nothing but mathematical structures and, as such, would support ESR, not traditional scientific realism.

In sum, Psillos' assumptions do not support the claim that specific non-structural components survive theory change.[14] What survives of the three assumptions appears to be thoroughly structural. Yet, even if we were to find clear cases of specific non-structural theoretical component preservation, we would still have to ask whether such components are essential in the prediction-making and explanatory aspects of theories. If they are not, their preservation is irrelevant for realist purposes.

Before we bring this section to a close, it is worth mulling over another one of Psillos' objections to ESR that alleges epistemic access to (potentially specific) non-structural components. In his own words:

> ... it isn't clear why the first-order properties of unobservable entities are unknowable. They are, after all, part and parcel of their causal role. So, if all these entities are individuated

[13] Our understanding of this relation is adjusted by the factor $\frac{1}{2}$.

[14] Redhead makes a similar observation (without however elaborating) when he says: 'Psillos presents detailed case studies for the examples of caloric and ether but what the discussion boils down to seems to be that structural aspects of the old theory are preserved in the new theory' [20, p. 344].

> and become known via their causal role, there is no reason to think that their first-order properties, though contributing to causal role, are unknowable [17, p. 17]; see also his [16, p. 156].
>
> ... these *in re* structures are individuated by their nonstructural properties since it's in virtue of these (nonstructural) properties that they have causal unity and are distinguished from other *in re* structures [18, p. 567].

In other words, how can we claim to know the causal role of entities without knowing their (potentially specific) non-structural properties? Following Grover Maxwell, Psillos equates non-structural properties with first-order properties. Maxwell's reason for this identification seems to be Russell's idea that the non-structural properties of percepts need not resemble the non-structural properties of their external world causes. Yet Maxwell's identification is unwarranted. Non-structural properties (specific or otherwise) need not be restricted to first-order properties in Russell's system. Moreover, Maxwell's idea is certainly not a consequence of his accepting the Ramsey-sentence approach. The Ramsey-sentence existentially quantifies over all theoretical properties regardless of whether these are first- or higher-order. It thus does not force its advocates to espouse an epistemic distinction between first-order and higher-order theoretical properties.

Psillos' (and Maxwell's) misconstrual notwithstanding, the question still stands: Can we know the causal role of entities without knowing their (potentially specific) non-structural properties? The answer to this question is 'Yes'. ESR does not deny that the specific non-structural properties of objects play an integral (and perhaps even necessary) causal role. Rather it holds that we have limited access to these properties, i.e. we can only know them up to isomorphism. Being necessary for a causal role does not equal being epistemically accessible. To illustrate this point consider the following analogy. Suppose you have been mugged but you don't exactly know by whom. Suppose further that unbeknownst to you the assailant mugged you because he was necessarily evil—a specific non-structural property he possesses. Do you need to know this property to know that somebody mugged you? Of course not! Likewise in the case at hand, we need not know the (potentially specific) non-structural properties of causes in order to know something about the causes. Indeed, if the epistemic structural realist is right, it is simply not possible to know specific non-structural properties.

At times Psillos' reasoning comes across as an instance of *argumentum ad consequentiam*. It starts with the premise that epistemic access to the specific non-structural properties of unobservables guarantees that our knowledge is realist. It then adds the premise that it is desirable for our knowledge to be realist. From this it is concluded that we have epistemic access to the specific non-structural properties of unobservables. It goes without saying that whether or not we have epistemic access to specific non-structural properties cannot be decided by what would be enough to save us from collapse to an unwanted form of realism or even anti-realism.

I would like to end this section with a challenge to traditional scientific realists. The challenge is quite simple. Identify one specific non-structural component that: (i) plays an essential role in the predictive and explanatory success of an abandoned theory, (ii) has survived into that theory's successor theories and (iii) cannot be replaced by a structurally identical analogue. Accomplish that and in one swift stroke ESR will be rendered lifeless.

3.6 Turning the Tables Around

Early on in our investigation I asserted that the two collapse claims are best understood in the form of a dilemma: Someone who wants to support ESR can plump for either a pure version that collapses to an exceedingly weak form of realism or an impure version that collapses to traditional realism. Over the course of this investigation, I called into question the second disjunct of this dilemma, arguing that impure versions of ESR do not automatically collapse to traditional realism. I have not called into question the first disjunct because I believe, as most epistemic structural realists do, that it contains a kernel of truth. Those who advocate pure ESR willingly understand it to be a weak, perhaps even a very weak, form of realism. After all, it was part of the original marketing strategy of the position to straddle the space between traditional scientific realism and empiricist anti-realism, i.e. making assertions that are weaker than those made by the former but stronger than those made by the latter. Telling epistemic structural realists that their view is a weak form of realism is therefore not an objection but an unnecessary reminder.

A more delectable upshot of this whole discussion is that we can now turn the tables on the traditional scientific realists by presenting them with an unpleasant dilemma: Either insist on specific non-structural knowledge of unobservables but then show up empty-handed (if the above challenge remains unmet, as I believe it will) and hence render your view false or drop the claim to specific non-structural knowledge but then experience a collapse to some form of ESR. Put bluntly, submit or perish!

I spent a good deal of energy in Sections 3.4 and 3.5 above trying to convey the idea that the traditional scientific realist opts for the first disjunct of the current dilemma. The truth of the matter is that this has not always been the case. Plenty of scientific realists have over the years expressed views that at the very least bear a striking similarity to ESR. In a seminal article on scientific realism, for example, Ernan McMullin emphasises the motivational importance of the convergence of structural explanations in the history of science. He asserts that '[i]t is, in part at least, because the history of science testifies to a substantial continuity in theoretical structures that we are led to the doctrine of scientific realism at all' [14, p. 22]. Similarly, Howard Stein has this to say: 'our science comes closest to comprehending "the real", not in its account of "substances" and their kinds, but in its account of the "Forms" which phenomena 'imitate' (for "Forms" read 'theoretical structures', for "imitate", "are represented by")' [24, p. 57]. Even Psillos, the arch-enemy of the structural realist, can at times be read this way. In the passage quoted earlier where the threat of ESR's collapse to traditional scientific realism looms he states: 'to say what an entity *is* is to show *how this entity is structured*' [16, p. 156], [emphasis in original]. And he adds '[a]n exhaustive specification of this set of properties and relations leaves nothing left out' [16, p. 156]. I am sure the reader will appreciate the irony here as this claim betrays a collapse that is the inverse of what its author originally envisaged.

Psillos will surely protest that by 'structural specification' he does not mean the same thing as the epistemic structural realists. For him this specification involves concrete structures whereas for epistemic structural realists it involves abstract structures. Even so, to demonstrate how a specific entity or system is structured requires nothing more than a specification of its abstract structure. It is not knowledge of the elusive specific non-structural components that allows us to assert the claim that 'it is *this* (as opposed to *that*) entity that is so structured' but the context—causal-perceptual in mine and Russell's view—in which it is uttered. Thus even though Coulomb's law of electrostatics and Newton's law of gravity are structurally identical, the context permits a different empirical interpretation of the quantities involved, e.g. we measure mass via instruments like the triple beam balance and charge via instruments like the electrometer.

3.7 Correspondence Without Reference?

I would like to end this paper by reflecting on a more radical reading of ESR. Let us first go back to the basics of scientific realism. What makes a view realist? Putnam states two conditions, which many realists endorse and which he attributes to Boyd: '(1) Terms in a mature science typically refer. (2) The laws of a theory belonging to a mature science are typically approximately true.' [19, p. 179]. Some scientific realists push for a stronger reading of (1), according to which, the successful reference of a theory's (observational and theoretical) terms is a necessary condition for that theory's approximate truth. This assumption has landed scientific realists into hot water. Laudan [12] famously takes advantage of the posited relationship between successful reference and approximate truth to argue against realism. To be exact, he argues that since nowadays we consider the central terms of empirically and explanatorily successful past theories to be non-referential we can no longer claim that their respective theories are approximately true. Recall that many realists want to preserve inferences from the empirical and explanatory success of theories to their approximate truth. Laudan's argument throws a spanner in the works of such inferences.

One realist reaction to Laudan has been to deny the view that reference is a necessary condition for approximate truth. To make this point Hardin and Rosenberg [9] offer a case from the history of biology. They claim that even though there is nothing in Mendel's 1866 theory that corresponds to our concept of a gene, the theory contains some important truths and can therefore be thought of as approximately true (p. 606). Hardin and Rosenberg's defence of scientific realism does not rely solely on the severance of the allegedly necessary connection between successful reference and approximate truth. Their approach is multifaceted and includes the deployment of causally-oriented theories of reference. For instance, they offer an alternative explanation of the Mendel case, according to which Mendel's central theoretical terms do in fact refer (in the causal-historical sense) regardless of the incorrect descriptive content associated with them. More generally, Hardin and Rosenberg rule that 'referential successes [must] be judged on a case by case basis' (p. 608).[15]

A more radical realist reaction to Laudan's challenge has recently been made by Cruse and Papineau [5]. According to them, the cognitively significant content of a scientific theory, i.e. what the scientific theory is really and meaningfully about, is captured by its Ramsey sentence. Since the Ramsey sentence of a theory turns theoretical predicates into existentially quantified variables, such variables presumably cannot be said to refer to any particular object. Cruse and Papineau take this to mean that 'the referential status of theoretical terms is irrelevant' (p. 174). In their view, the question whether successful theoretical term reference and approximate truth are correlated does not even arise. To understand how this view is more radical than that of Hardin and Rosenberg we need only consider that the denial of the necessary link between referential success and approximate truth leaves open the door that the two notions are highly correlated.

In a similar vein, Worrall has in the last few years rejected referential semantics, opting for a provocative interpretation of Ramsey-style ESR.

> If it is assumed that to be a 'real realist' one must assert that the terms in our current theories refer as part of an acceptance of a correspondence or semantic view of truth as the account of what it means for our theories to have latched on to the real structure of the world, and it is assumed that the realist must develop some sort of weakened version of correspondence as her account of 'approximate correspondence with reality' then [E]SR does not count as 'real realism'... But there is no reason why the way in which a theory mirrors reality should be the usual term-by-term mapping described by traditional semantics. Indeed, as I have remarked several times already, if we are talking about an epistemically accessible notion then it cannot be! [E]SR in fact takes it that the mathematical structure of a theory may globally reflect reality without each of its components necessarily referring to a separate item of that reality [31, pp. 32–33].

In my view there are two solid reasons to dismiss this approach. The first concerns Worrall's (as well as Cruse and Papineau's) incoherent conception of the Ramsey sentence. Though it is true that the variables in Ramsified theories do not range over particular objects or properties, it is also true that they range over sets of such objects and properties. Thus Cruse, Papineau and Worrall might be warranted to infer that theoretical terms do not refer to singular objects/properties but they are not similarly warranted to infer that no non-global reference takes place whatsoever. After all, it seems that we are fully capable of referring to sets of objects and we do so all the time regardless of whether the sets contain observables or unobservables.[16]

The second reason concerns Cruse, Papineau and Worrall's incoherent use of the notion of approximate truth. To the extent that a theory can be approximately

---

[15] In my view, the realist must choose on some principled basis which theory of reference to apply, otherwise the whole issue becomes trivialised.

[16] Along similar lines, Grover Maxwell [13] has argued that the theoretical variables of a Ramsey sentence refer indirectly to unobservable objects. They do so implicitly via their logical relations to unRamsified (i.e. observational) terms that refer directly to observable objects (pp. 182–183).

true with respect to the unobservable world it is surely telling us something about how the unobservable world is structured. But how can we attribute structure to the unobservable world without saying something about its entities, their properties and relations? Under the traditional conception of the correspondence theory of truth, a scientific theory's truth or approximate truth implies that the theory's terms refer, among other things, to unobservables.[17] Under Worrall's conception, we are asked to imagine that the structure of our theory globally reflects the structure of the unobservable reality. But what does it mean for a theory's structure to *globally reflect* the (structure of the) world? Without an unambiguous semantics that tells us under what conditions such a structure truly or falsely ascribes features to the unobservable world, Worrall's proposal cannot be properly evaluated.[18]

Those who found Hardin and Rosenberg's more modest approach compelling may be unsympathetic to my second objection. After all, does not the denial of the necessary connection between referential success and approximate truth also not entail the possibility that we can have approximately true statements whose terms do not succeed in referring? In my view, it does not! Hardin and Rosenberg specifically target the central terms of scientific theories. Otherwise put, they deny that the reference of central theoretical terms is a necessary condition for that theory's approximate truth. Whether the theory's approximate truth or, better yet, approximately true parts can be assessed without the reference of at least some of the theory's terms is something they leave unanswered. On the basis of their examples, there is in fact good reason to believe that we cannot have approximate truth without referential semantics. Take Mendel's case again. His theory may not contain anything corresponding to our concept of a gene but, in so far as it is true, it contains terms that we take to refer even today, namely hereditary factors that play the role of the unobservable causes of phenotypic traits. Unless a clear case can be made that claims about the unobservable world can be approximately true without at the same time the terms appearing in *those* claims being referential, Worrall's correspondence-without-reference suggestion remains just another flight of fancy.

3.8 Conclusion

Traditional scientific realism cannot be upheld if its advocates: (a) insist on a type of knowledge (i.e. specific non-structural knowledge) that cannot be substantiated or (b) subscribe to a 'purely' structural account of the world. Structural realists ought to encourage their old-fashioned realist brothers and sisters to come out of the closet and embrace their true identity.

---

[17] To establish that approximate truth is a sufficient condition for referential success is of course to establish that referential success is a necessary condition for approximate truth.

[18] Even then, the advocate of this approach must still explain why it is that referential semantics is good for observational terms but bad otherwise.

Acknowledgments I am very grateful to Elaine Landry for inviting me to the *Structure, Objects and Causality* workshop, which she masterfully organised in Banff, Canada in August 2007 and where this paper was first presented. I am also very grateful to the participants of the workshop as well as to my colleague Ludwig Fahrbach for providing valuable feedback. Finally, I am indebted to the German Research Foundation (Deutsche Forschungsgemeinschaft) for making the writing of this paper possible by funding my research (project B6 of the interdisciplinary research unit FOR 600 'Functional Concepts and Frames').

References

1. Bain, J. (1998) *Representations of Spacetime: Formalism and Ontological Commitment.* Unpublished PhD thesis. University of Pittsburgh.
2. Bradie, M.P. (1977) The Development of Russell's Structural Postulates. *Philosophy of Science* **44**: 441–463.
3. Brading, K. and E. Landry (2006) Scientific Structuralism: Presentation and Representation. *Philosophy of Science*. **73**: 571–581.
4. Craig, E., ed. (1998) *Routledge Encyclopaedia of Philosophy,* Vol. 8. London and New York: Routledge.
5. Cruse, P. and D. Papineau (2002) Scientific Realism without Reference. In M. Marsonet (ed.), *The Problem of Realism* (pp. 174–189). Aldershot: Ashgate.
6. d' Alembert, Jean Le Rond. ([1757] 2006) Conservation des *Forces* Vives. In *Encyclopédie ou Dictionnaire Raisonné des Sciences, des Arts et des Métiers,* Vol. 7 (pp. 114–116). Translated by John S.D. Glaus for *The Encyclopædia of Diderot & d'Alembert Collaborative Translation Project,* Ann Arbor MI: Scholarly Publishing Office of the University of Michigan Library.
7. French, S. and D. Krause (2006) *Identity in Physics: A Historical Philosophical, and Formal Analysis*. Oxford: Clarendon Press.
8. Frigg, R. and I. Votsis (2011) Everything you Always Wanted to Know about Structural Realism but Were Afraid to Ask. *European Journal for the Philosophy of Science* **1**: 227–276.
9. Hardin, C.L. and Rosenberg, A. (1982) In Defence of Convergent Realism. *Philosophy of Science* **49**: 604–615.
10. Ketland, J. (2004) Empirical Adequacy and Ramsification. *British Journal for the Philosophy of Science* **55**(2): 287–300.
11. Ladyman, J. (1998) What is Structural Realism? *Studies in History and Philosophy of Science* **29**: 409–424.
12. Laudan, L. (1981) A Confutation of Convergent Realism. *Philosophy of Science* **48**: 19–49.
13. Maxwell, G. (1970) Structural Realism and the Meaning of Theoretical Terms. In S. Winokur and M. Radner (eds.), *Analyses of Theories, and Methods of Physics and Psychology* (pp. 181–192). Minneapolis MN: University of Minnesota Press.
14. McMullin, E. (1984) A Case for Scientific Realism. In J. Leplin (ed.), *Scientific Realism* (pp. 8–40). Berkeley CA: University of California Press.
15. Papineau, D. (1996) Introduction. In D. Papineau (ed.), *The Philosophy of Science* (pp. 1–20). Oxford: Oxford University Press.
16. Psillos, S. (1999) *Scientific Realism: How Science Tracks Truth.* London: Routledge.
17. Psillos, S. (2001) Is Structural Realism Possible? *Philosophy of Science* **68**: 13–24.
18. Psillos, S. (2006) The Structure, the Whole Structure and Nothing but the Structure? *Philosophy of Science* **73**(5): 560–570.
19. Putnam, H. (1975) What Is "Realism"? *Proceedings of the Aristotelian Society* **76**: 177–194.
20. Redhead, M.L.G. (2001) Quests of a Realist, Review Article of Stathis Psillos's Scientific Realism: How Science Tracks Truth. *Metascience* **10**: 341–347.
21. Russell, B. (1927) *The Analysis of Matter*. London: George Allen and Unwin.
22. Russell, B. ([1948] 2009) *Human Knowledge: Its Scope and Limits*. London: George Allen and Unwin.

23. Russell, B. ([1968] 1998) *The Autobiography of Bertrand Russell,* Vol. 2. London: George Allen and Unwin.
24. Stein, H. (1989) Yes, But... Some Skeptical Remarks on Realism and Anti-realism. *Dialectica* **43**(1–2): 47–65.
25. Votsis, I. (2003) Is Structure Not Enough? *Philosophy of Science* **70**(5): 879–890.
26. Votsis, I. (2004) *The Epistemological Status of Scientific Theories: An Investigation of the Structural Realist Account*. Unpublished PhD Thesis. London School of Economics.
27. Votsis, I. (2005) The Upward Path to Structural Realism. *Philosophy of Science* **72**(5): 1361–1372.
28. Votsis, I. (2007) Uninterpreted Equations and the Structure-Nature Distinction. *Philosophical Inquiry* **29**(1–2): 57–71.
29. Votsis, I. (2011) Structural Realism: Continuity and its Limits. In P. Bokulich and A. Bokulich, (eds.) *Scientific Structuralism* (pp. 105–117). Dordrecht: Springer.
30. Worrall, J. (1989) Structural Realism: The Best of Both Worlds? In D. Papineau, (ed.) *The Philosophy of Science* (pp. 139–165). Oxford: Oxford University Press, 1996.
31. Worrall, J. (2007) Miracles and Models: Why Reports of the Death of Structural Realism May Be Exaggerated. *Royal Institute of Philosophy Supplement* **82**: 125–154.
32. Worrall, J. (2011) Underdetermination, Realism and Empirical Equivalence. *Synthèse* **180**(2): 157–172.
33. Worrall, J. and E. Zahar, (2001) Ramseyfication and Structural Realism. Appendix IV in Zahar E. *Poincaré's Philosophy: From Conventionalism to Phenomenology*. Chicago and La Salle, IL: Open Court.

Chapter 4
Miracles and Structural Realism

John Worrall

4.1 Introduction

The often breathtaking predictive success of *some* theories in contemporary science inclines most of us towards scientific realism: surely what those theories say about the 'unseen world' lying 'beyond the phenomena' must be at least approximately correct if they can score such dramatic, empirically checkable, successes? The facts about theory-change in science, on the other hand, seem to speak in favour of an anti-realist view: scientists have in the past held theories that were also dramatically predictively successful and yet which are now 'known to be false' (because they are inconsistent with our latest theories). Given this, what guarantee can there possibly be that our latest theories will not themselves be rejected and replaced by quite different ones at some time in the future? And if so, how can we reasonably hold that our current theories are true? And when we think about how radical some of those theory-changes appear to have been, how can we hold that our current theories are likely even to be *approximately* true? My [21] argued that, although these two much-discussed considerations thus seem to pull sharply in opposite directions, they can in fact be reconciled within a version of realism—namely, *structural realism*.

The first, apparently pro-realist, consideration has often been developed as 'the no-miracles argument' (hereafter the NMA, or rather, as we shall see, 'the' NMA). The intuition is roughly that it would be a miracle if current scientific theories enjoyed the predictive successes that they do if what they claimed was going on 'behind' the phenomena were not at least approximately correct; but we should not accept that miracles have occurred unless there is no non-miraculous alternative; and here the (approximate) truth of what the theories say about the 'noumenal' world is exactly a non-miraculous alternative explanation of their empirical success. The second, apparently anti-realist, consideration has often been developed as 'the pessimistic (meta-) induction' (hereafter, the PI). Roughly: theories that were accepted in the past (exactly on the basis of the predictive success emphasised by the

J. Worrall (✉)
Philosophy, Logic & Scientific Method, London School of Economics, London, UK
e-mail: j.worrall@lse.ac.uk

E.M. Landry, D.P. Rickles (eds.), *Structural Realism*, The Western Ontario Series in Philosophy of Science 77, DOI 10.1007/978-94-007-2579-9_4,

NMA) have subsequently turned out to be (perhaps radically) false; so, we should infer (inductively/probabilistically) that our current theories are (perhaps radically) false too.

However, Colin Howson has argued (see [6], chapter 3) that the NMA—in so far as it can be turned into a precise argument at all—in fact embodies an elementary probabilistic fallacy: often called 'the base rate fallacy'. While, ironically enough, Peter Lewis [11] has (independently) argued that essentially the same fallacy underlies, and therefore vitiates, its seeming-competitor argument—the PI.

If these arguments really do establish that the NMA and the PI are fallacious, then this would seem to destroy the basic problematic at which structural realism is addressed. And Magnus and Callender have indeed recently urged that, since 'the major considerations for and against realism come to naught', the whole scientific realism debate (at any rate in what they call the 'wholesale' sense) should be 'dissolved', as resulting in nothing but 'ennui', [14, pp. 321–322].

Is there anything in these recent arguments that should concern the structural realist or force her into a state of terminal ennui? In this paper I restrict myself to the concerns about 'the' NMA (leaving those about 'the' PI for another occasion [26]). I argue that the 'difficulties' raised in the literature are no more than artefacts of the (misguided) way in which the considerations underlying 'the' no miracles argument have been formalised, or 'modelled', as precise arguments. The underlying intuition remains untouched and remains a good (though, of course, far from conclusive) reason for adopting structural realism.[1]

4.2 No Miracles Reconsidered: The Intuitions

Consider a classic, and by now well-worn, example that elicits the intuitive 'no miracles response' (at any rate in yours truly). Fresnel's theory states that light consists of (not directly observable) waves transmitted through a (not directly observable) all-pervading elastic medium. His theory turned out to entail, as Poisson demonstrated but as Fresnel himself had never suspected, the directly empirically checkable result that if a small opaque disc is held in light diverging from a point source and if the 'geometric shadow' of the disc (that is, the area of complete darkness that would exist if the laws of geometric optics were strictly correct) is carefully examined, then the centre of that 'shadow' will in fact be seen to be illuminated, and indeed just as strongly illuminated as if no opaque disc were present. Although most of the French Academicians thought that this was a clear-cut *reductio* of the theory, when

[1] There have also been a number of direct criticisms of Structural Realism in the recent literature, many of them based on what might be called the 'Newman objection' (see [15], the revival of that argument in [2], and more recently [9]). These criticisms are also not dealt with in the present paper but *are* addressed and rebutted in [25].

Arago performed the experiment it turned out that the 'white spot' does indeed, and contrary to all prior expectations, exist.[2]

Whatever esoteric philosophical considerations may be raised, it is difficult to resist the feeling that if a theory can make such a striking, seemingly improbable prediction that nonetheless turns out to be empirically correct, then the theory must somehow be 'approximately true'—it must have somehow latched on, no doubt in an approximate (but nonetheless substantial) way, to the 'deep structure' of the universe: to how things really are in the 'noumenal world' behind or beyond the phenomena. Duhem, who was not the instrumentalist he is often considered but rather a structural realist, put it eloquently [3, p. 28]:

> The highest test, therefore, of our holding a classification as a natural one is to ask it to indicate in advance things which the future alone will reveal. And when the experiment is made and confirms the predictions obtained from our theory, we feel strengthened in our conviction that the relations established by our reason among abstract notions truly correspond to relations among things.

A theory gives us a 'natural classification', according to Duhem, just in case 'the relations' it posits 'truly correspond to relations among things'. Our 'conviction' that Fresnel's theory represents such a natural classification is 'strengthened' because it would, it seems, be extremely unlikely that the theory could have such a striking, and empirically correct consequence (that it should 'indicate in advance things which the future alone [revealed]'), were what it said about the reality that underlies phenomena such as the 'white spot' *not* a natural classification, that is, not in some sort of (approximate) correspondence with reality.

Or take another much-discussed example: Quantum Electrodynamics predicts the magnetic moment of the electron to better than one part in a billion. (Theory yields $1.00115965246 \pm 0.0000000002$ as the value, while, early in this century, the most sensitive observations yielded the value of $1.00115965221 \pm 0.00000000003$!).[3] Again it seems difficult to resist the feeling that the theory *must* somehow have latched on to the way things are 'underneath' the phenomena if it can get such a prediction correct to such an implausibly high degree of accuracy.

The talk about theories being 'approximately true' or (better) 'somehow latching on to the way that things are beneath the phenomena' may be imprecise but it is clearly necessary. This is emphasised by the considerations underlying the PI—later (better) theories tell us that Fresnel's theory is strictly false—but the point is independent of that argument. No one believes, even ahead of any further 'scientific revolution', that Quantum Field Theory is true—indeed there are questions about whether a fully coherent theory can be articulated at the present time. No one even believes (or ought to believe) that Quantum Mechanics itself, for all its stunning success, will survive entirely unscathed. (Its two basic postulates are clearly mutu-

[2] The real history, as I show in my [20], was a good deal more interesting and a great deal less clear-cut. However the real historical details, although they do centrally affect the issue of what counts as a successful prediction (and why predictions carry more confirmatory weight), do *not* affect the philosophical issue about the link between successful prediction and realism.

[3] Values quoted from [6].

ally incoherent; and it also fails to cohere with the General Theory of Relativity.) So the realist claim must in general be that it would be a 'miracle', not if the theory at issue failed to be 'outright' true, but rather if it failed to be somehow *approximately* true.

Whether the NMA can be given some more exact construal will be our central topic. But it does seem clear that science centrally embodies the underlying intuition and would not be possible if it did not. This is reflected in the fact that appeal to the intuition is implicit, not just in the justification of observation-transcendent theories, but also in the justification of the standard empirical generalisations that everyone—including those who are anti-realist about scientific theories—accepts. This has gone largely unacknowledged in the recent literature, but it *was* recognised, and emphasised, by Poincaré, who wrote [16, pp. 149–150]:

> We have verified a simple law in a considerable number of particular cases. We refuse to admit that this coincidence, so often repeated, is a result of mere chance and we conclude that the law must be true in the general case.
>
> Kepler remarks that the positions of the planets observed by Tycho are all on the same ellipse. Not for one moment does he think that, by a singular freak of chance, Tycho had never looked at the heavens except at the very moment when the path of the planet happened to cut that ellipse ... [I]f a simple law has been observed in several particular cases, we may legitimately suppose that it will be true in analogous cases. To refuse to admit this would be to attribute an inadmissible role to chance.

It would, Poincaré is saying, constitute an incredible coincidence—a 'miracle', if you like—if Kepler's simple (first) law were instantiated by all the planetary positions that had so far been checked but was not in general true (that is, not instantiated also by all the—past and future—unobserved planetary positions). No instrumentalist or constructive empiricist known to me fails to endorse the acceptance as rational of standard empirical generalisations (again generally, as the Kepler case illustrates, as approximately, rather than outright, true)—on the basis of what is of course bound to be a finite set of actual observations. They are therefore all relying—as Poincaré argues—on exactly the no miracles consideration that they vigorously deny should be thought of as persuasive when it comes to observation-'transcendent', theoretical claims.[4]

The intuition underlying the no miracles argument also underwrites persuasive arguments in a variety *both* of scientific *and* of more commonplace circumstances. Maxwell's work initially left open the possibility that there might be two different media filling space: the optical ether *and* the electromagnetic field. But once he had discovered that waves were transmitted through the field at the velocity of light, he immediately inferred that it would be miraculous if there were two media each of which just happened to transmit disturbances at exactly the same rate; and hence he inferred that there is only one medium—the field—and that light is in fact an electromagnetic wave. Einstein refused to admit that the parameter measuring a

[4] This is why it seems disingenuous of Magnus and Callender [14] to take it that everyone agrees that we are entitled to make the standard 'horizontal' inductive inferences and then to lay this aside in their discussions of differences of opinion concerning the NMA.

body's responsiveness to an applied force (its inertial mass) and its gravitational action (its gravitational mass) could be identical by accident. It would be a 'miracle' if these two conceptually distinct quantities just happened invariably to have the same value—so some non-miraculous account must be sought and was of course found in the form of the General Theory of Relativity.

Admittedly it is easy to produce alleged 'miraculous coincidences' pretty well at will and many people have been seduced by cooked-up 'coincidences' into accepting conclusions that are themselves quite staggeringly improbable—I think in particular of arguments concerning the so-called anthropic principle and some arguments for the existence of god. So we certainly need to take care in this area. Nonetheless it seems difficult to resist the idea that there is *something* important in the intuitions underlying the truly persuasive instances of the NMA.

Philosophers—on the whole quite rightly—are, however, suspicious of 'intuitions' and try to capture what, if anything, is valuable in them in more rigorous arguments, whose credentials can in turn be examined more sharply. So how, if at all, can the intuitions elicited by the predictive success of at least some theories be captured in some more precise argument?

4.3 How *Not* to Formalise the NMA

4.3.1 The Scope of 'the' Argument

What exact scope should we expect such a formalised argument to have? Hilary Putnam [18, p. 19], suggested that we should think of scientific realism as itself a sort of 'overarching scientific hypothesis' that (allegedly) provides the 'best explanation' of the success of 'science'. This idea was developed by Richard Boyd [1] and endorsed by Stathis Psillos [17], who calls it the 'explanationist defence' of scientific realism. So the idea is that the argument is to be regarded as a grand, meta-level 'abduction' or 'inference to the best explanation'. The best (perhaps only) explanation of the success of science in general is the truth (or approximate truth) of its theories. So we are entitled to infer that its theories are indeed (at least approximately) true—that is, we are entitled to infer the thesis of scientific realism. And this inference is a scientific inference, no different in form from the particular inferences to the best explanation routinely drawn within science.

But surely a number of fundamental objections to this form of the argument were (or ought to have been) apparent from the beginning. The underlying idea is that there is some general scientific method for producing theories—one that has been so successful that we are entitled to infer at least to the approximate truth of its products. But, allowing for the moment that there may be such a method, it is certainly not uniformly successful. There is a wide variety of sciences, not all of them 'surprisingly successful', certainly not in any sense that elicits the 'no miracles intuition'. Nothing known to me in the sciences of sociology, parts of psychology, dietetics etc provides any reason to make one think that their theories have successfully penetrated to the noumenal world ‘beyond’ the phenomena. Nor does the fact that, for example, ‘scientific’ Creationism has successfully continued to attract adherents and exercise a good deal of political and social power have the slightest implication for the correctness of its theoretical claims. Nor even does the sociological fact that some theory has attracted a good deal of reasonably extended attention within ‘respectable’ scientific circles present any temptation in itself to cry ‘no miracles’. It does seem that Larry Laudan, for example, must have been using a far-too-undemanding notion of what it takes for a theory to be ‘successful’ when including such vapid ideas as the aetherial musings of Hartley and LeSage amongst those theories that ‘were once-successful but are now known to be [radically] false’ [10].

Now Putnam was of course quick to restrict the argument, from the beginning, to ‘mature’ science. But, as I have argued elsewhere [21], the only reasonable characterisation seems to be that a science achieves ‘maturity’ once its basic theories have turned out to be predictively successful. So this means in fact that we must first check whether the theories in some field have enjoyed predictive success before including them within the scope of the ‘overall’ inference. But this surely means in turn that the alleged general character of the argument evaporates on reflection: the argument is general only in so far as it consists of the union of a series of individual arguments about particular theories—arguments that infer the likely approximate truth of individual theories such as Fresnel’s or Quantum Field Theory from their particular empirical successes. Aside from this, how could any general inference to the best explanation of the success of science ever have been thought of as itself a *scientific* explanation?

There has been a good deal of talk in the explanation literature about ‘loveliness’ and the like[5]—the idea that explanatoriness is a quality that a theory may possess over and above its degree of empirical support. However, the history of science seems to me to show conclusively that which theories are found to be ‘explanatory’ is a historically contingent issue dependent on which theories have the highest degree of support. A classic instance involves action at a distance—initially branded as incomprehensible and so no possible part of any theory that could count as ‘explanatory’ (Newton himself of course shared this view), it became a perfectly acceptable, ‘explanatory’ notion because of the overwhelming empirical success of Newton’s theory: to the extent that when, for example, Coulomb came to formulate his law of electrostatic attraction and repulsion, there was no concern at all about its being an action at a distance theory. Of course later, in the light of the still better empirically supported theories of relativity which reject the idea of action at a distance, it was again abandoned as a ‘non-explanatory’ idea. What we find ‘explanatory’ or not is dependent on empirical success. And this means independent predictive success.

[5] See for example [13].

If scientific realism were to be considered an overarching meta-level but nonetheless scientific explanation of science's success, it would therefore, need to have *evidence* in its favour. What could that be? What, more specifically, could count as an independent predictive test of (any version of) scientific realism as a general, in Magnus and Callender's terms [14] 'wholesale', view? The only candidate would seem to be the (meta-level) 'prediction' that the next theory to be accepted in any field will be successful, that is, will enjoy independent predictive success over and above that enjoyed by the currently accepted theory. But, restricting consideration, as we already saw we must, to mature science, this will be trivially or definitionally true; and so its fulfilment cannot be regarded as an 'empirical success' on a par with those enjoyed by (ordinary 'object level') scientific theories themselves. A field becomes mature when its accepted theories are successfully predictive, and science would clearly never accept a new theory as superior to a currently accepted one unless that new theory not only scored the same successes as its predecessor, but also enjoyed predictive successes over and above those shared with the one it displaced.

This is all ahead of the obvious, though nonetheless telling, objection, made by Arthur Fine [4], that any such 'explanationist defence' of scientific realism would be circular. I would put the point as follows. If someone held that it was reasonable to infer to the (approximate) truth of our 'most explanatory' (really best empirically supported) theories, then she would already be a scientific realist—albeit via what I think is the correct route through the union of a series of inferences about particular successful theories, rather than as the result of some fancy meta-level abduction about 'science in general'. For then she would have no need to cite any such meta-level abduction. On the other hand if someone were to question 'inference to the best explanation' (inference to the best supported theory) as regards particular scientific theories, then in consistency she could not fail to question this alleged grand meta-level 'inference to the best explanation' which *at best* (that is laying aside the above objections) has the same logical form.[6]

In sum, then, it surely always was a mistake to think of any 'wholesale' version of scientific realism as a sort of general inference to the best explanation. What are successful or not, what elicit the no miracles intuition or not, are particular *individual* theories—such as Fresnel's wave theory of light or Quantum Field Theory. In so far as there is any sort of 'wholesale' case to be made for scientific realism it is simply as the union of a whole set of specific cases for individual theories.[7] Moreover any

---

[6] Psillos in particular has attempted to defuse Fine's argument; but the attempt to argue in effect that some circles are unproblematic seems to me deeply unconvincing. (See my [22] response to [12] which essays a structurally identical argument.)

[7] There is a fundamental incoherence in the Magnus and Callender paper [14]. While dismissing the 'wholesale' argument for realism, partly on the grounds that the NMA is fallacious, they applaud investigation of 'retail' realist arguments—for particular theories or particular entities (such as 'the' atom). The problem is, of course, that the 'retail' arguments, as suggested above, can all be construed (and ultimately only construed) as instances of some form or other of the NMA. We believe in atoms, because atomic theories have had striking empirical successes to an

such case is made for a *philosophical* thesis not in any sense a 'scientific' one, since there is, and can be, no question of that thesis itself scoring any independent predictive empirical success.

4.3.2 'Retail' Realist Arguments: (Objective) Probabilities Won't Help

Any sensible version of the NMA, then, will be of the 'retail' variety in that its conclusion will be that it is reasonable to hold that some *particular* theory—the wave theory of light, GTR, QFT, . . . —is approximately true. Moreover the success involved in the premise of any sensible version will not be any vague, generic, 'wholesale' notion of success but the genuine *predictive* success of the particular theory at issue: the theory must make a prediction of a general kind of empirical result, one that corresponds to the outcomes of observations or experiments. 'Prediction' here, as I have explained elsewhere (see in particular my [23]), need not involve novel, that is, hitherto undiscovered phenomena. The operative condition is that the general phenomenon must not have been 'used in the construction' of the theory at issue (obviously this will automatically be satisfied by any piece of 'new' evidence that was unsuspected at the time when the theory concerned was first formulated). No one is going to exclaim when confronted, say, with some version of Ptolemaic geocentric theory that correctly entails that the planets exhibit stations and retrogressions 'Wow! That must mean that there is something about the theory's fundamental claims that must be at least approximately correct, otherwise it would be a miracle if it succeeded with such a striking prediction'. This is because there is a much more homely explanation of its 'success': parameters in the general Ptolemaic theory (relating sizes of epicycles and deferents, and the relative epicyclic and deferential velocities) had been fixed precisely on the basis of the previous observation of planetary stations and retrogressions, so that the particular version of Ptolemaic theory with parameters fixed in this way was *bound* to yield the phenomena at issue, irrespective of whether or not the overall theory of which it is a part has 'latched on to reality'. This demanding predictivist criterion of success rules out every theory in Laudan's 'plethora' of 'successful' theories that we allegedly now take to be radically false—with one exception: the 'classical' wave theory of light as a periodic disturbance in an elastic medium. Other theories on the list—such as the already mentioned gravitational and physiological ethers of Hartley and Lesage or the astronomical theory of the crystalline spheres—are surely classic instances of *ad hoc* theories. They identify an 'explanatory need'—how, for example, do the sun, planets and stars all move around the earth and why do they all orbit it in the same direction ?—but they 'solve' it (in the geocentric version of crystalline sphere theory by assuming that those astronomical objects are all embedded in concentric spheres

extent that seems entirely implausible if they are not 'on the right lines'. Maybe fancy ways of dressing up 'retail' realist arguments may disguise this fact, but it is a fact nonetheless.

that are themselves revolving in the same direction but at different rates about an axis passing through the Earth) without the slightest hint of any independent testability. The fact that a theory was taken seriously even by serious scientists is not something on which any sensible realist would rest any part of her case. Only predictive success counts.

At least at first blush, the impact of successful prediction for a specific individual theory T can be captured by the following informal argument. T has scored some spectacular predictive success; it would be a miracle if T could get such a phenomenon so exactly right if it were not itself at least approximately correct; but we should not accept that miracles have occurred if there is an alternative explanation of the state of affairs at issue; and there is exactly such a non-miraculous alternative in such cases—namely that T is at least approximately correct; hence we should infer that T is indeed approximately correct.[8]

If we are to capture this argument, and in particular the tricky notion of its being 'a miracle if T were to get evidence e correct without itself being "approximately" or "essentially" correct', in some more formal way, then surely the only realistic prospect is through a probabilistic reconstruction.[9] In investigating the prospects for such a reconstruction, let's first temporarily lay to one side the issues about approximation. The rather nebulous talk about it being a miracle if T had got such a phenomenon as e right if it were not true seems then to translate crisply into the assertion that the probability that e would happen were T false is extremely small: $p(e/\neg T) \approx 0$. And the fact that T (when taken together with accepted auxiliaries) deductively entails e 'translates' of course into the claim that $p(e/T) = 1$. Hence we have:

Premise 1 $p(e/T) = 1$ (e is entailed by T)
Premise 2 $p(e/\neg T) \approx 0$ (it would 'be a miracle if e had been the case were T not true')
Conclusion $p(T/e) \approx 1$ and hence, given that e has occurred, $p(T) \approx 1$.

There are, of course, entirely legitimate worries about what exactly the probabilities in these formulas mean, but laying these worries aside too for the moment (they will

[8] Although the claim that the approximate truth of T would explain its 'otherwise miraculous' success with some surprising prediction e sounds very plausible, it is by no means as obviously true as it might sound. Clearly if a theory is *true* then so are all its consequences—so if it entails some unlikely prediction that turns out to be correct, it seems reasonable to regard the theory's truth as the explanation of its success. But who has shown that all consequences of an 'approximately true' theory (or even, more restrictedly, all the empirically-checkable consequences of such are theory) must themselves be approximately true? Of course, if, as I recommend, 'approximately true' is taken, in structural realist manner, to amount to no more than 'will be retained, modulo the correspondence principle, in all further scientific theories' then this guarantee is supplied.

[9] The other alternative would be to construe the argument as some sort of (allegedly) formal 'Inference to the Best Explanation' which was not itself given a probabilistic construal. My reasons for rejecting this alternative are adumbrated later.

be re-raised very shortly), it is not difficult to show that, so long as they are indeed probabilities, then this reasoning is, as it stands, straightforwardly fallacious.

Here is a simple, and by now well-known, counterexample cited by Colin Howson [6, pp. 52–54]. Suppose that we have a diagnostic test for some disease D, and that this test (unfeasibly) has a zero rate of 'false negatives': that is, the probability of someone's testing negative if she does have the disease is equal to 0; and moreover a non-zero but (again unfeasibly) low 'false positive' rate: say, 1 in a 1000—that is, the probability of someone's testing positive even though they do not in fact have the disease is 1/1000. Suppose now that some particular person x has tested positive, what is the chance that x actually has the disease? In order to avoid changing terminology later, let T stand for the theory that x is suffering from D, while e stands for the evidential statement that x has produced a positive result in the diagnostic test at issue. The zero false negative rate is then just expressed by $p(e/T) = 1$; the low false positive rate by $p(e/\neg T) = 1/1000$; and the probability we are interested in, the probability of x's having the disease given that she has tested positive, is of course $p(T/e)$.

It is often asserted as an empirical result about human psychology (see, for example, [8] and [14]) that most people in these circumstances are inclined to infer from the fact that some person has tested positive and the fact that there is very little chance that x will test positive if she does not have D, that it is highly probable that she does have the disease. Such people would be reasoning in perfect agreement the above version of the NMA:

Premise 1 holds in the diagnostic case because x is certain to test positive (e) if she has the disease (T) (i.e. $p(e/T) = 1$); Premise 2 holds because it is extremely unlikely that x would test positive if she did not have the disease ($p(e/\neg T) = 1/1000 \approx 0$); and the conclusion being drawn is that the probability of x having the disease in view of the positive result—that is, $p(T/e)$—is very high.

Yet, as aficionados are well aware, this inference is an instance of the 'base rate fallacy'. Far from it following that the probability of T given e is very high, *any* non-extreme probability of T, given e, is in fact compatible with the truth of the two premises—even one that is arbitrarily close to zero. It all depends, of course, on the prior probability of T. In the diagnostic case we can, it seems, take that to mean the overall incidence of the disease. If the disease is very rare, a lot rarer than the rate of false positives, then the probability that x has the disease may be very low. So for example, if $p(T) = 10^{-6}$ then the probability here that the person who tested positive has the disease is, via a straightforward application of Bayes's theorem, only around 10^{-3}.

So our first stab at a probabilistic reconstruction of 'the' NMA produces a fallacy. Moreover, the prospects of producing a non-fallacious argument along these lines are surely not improved by reintroducing considerations of approximate, as opposed to 'outright', truth. As we saw, no sensible realist will want to claim anything stronger than that some theory T is approximately true, no matter how astounding its predictive success might have been. But modifying the claim in this way is not likely to help when it comes to reconstructing the NMA probabilistically.

Let $A(T)$ be the assertion that T is approximately true. The relationship between $A(T)$ and e is altogether less clear-cut than that between T and e. I am taking it that, the relevant auxiliaries being taken as given, T deductively entails e; but, on the other hand, $A(T)$, whatever it might precisely mean, presumably need not actually entail e. Nonetheless, since the aim of any version of the NMA is to have e have large impact on $A(T)$—to be reflected, if this reconstruction is to succeed, in an increase in $A(T)$'s probability once e has been observed— presumably the realist will need to claim that $p(e/A(T)) \approx 1$. And again the fundamental assumption here is that the evidence at issue would be very improbable were T not even approximately true, so the realist is presumably committed to the premise $p(e/\neg A(T)) \approx 0$. Hence we have a simple modification of the earlier argument:

Premise 1′ $p(e/A(T)) \approx 1$
Premise 2′ $p(e/\neg A(T)) \approx 0$
Conclusion $p(A(T)/e) \approx 1$; and hence, given that e has occurred, $p(A(T)) \approx 1$.

But then clearly the base rate problem kicks in just as before: depending on the value of the prior probability of $A(T)$ (the assertion that T is approximately true), any posterior for $A(T)$—including one as close to zero as you like—is compatible with the truth of premises 1′ and 2′.

If either of the above is the only or uniquely sensible way of capturing the intuitions underlying the NMA, then those intuitions must of course be abandoned entirely since there is no denying the fallaciousness of the base rate fallacy. It seems to me, however, not only that far from being the uniquely correct way to capture those intuitions, it should have been clear ahead of any analysis that no such reconstruction would work. The chief difficulty lies in the issues of how the relevant probabilities could possibly be interpreted in the case of Fresnel's theory or QFT or any other theory whose predictive success elicits the 'no miracles intuition'.

In the diagnostic case, the probabilities involved can arguably be interpreted as objective chances, reflecting —or perhaps constituted by—limiting relative frequencies: the test's false positive rate of 1 in 1000 reflects the assumption that *if* random selections from the whole population were continually made and the frequency recorded of those people who tested positive but failed to have the disease amongst all those testing positive, *then* that frequency would converge on 1/1000 as the number of selections increased indefinitely. Similarly the 'natural prior' in the diagnostic case is the overall incidence of the disease within the population: the proportion of those suffering from the disease is 1 in every million of population and hence if a series of selections were made at random from the population and the relative frequency of those having the disease recorded, then that frequency would converge on 1/1000000 as the number of selections increased indefinitely.[10]

[10] Notice, however, that this is hardly the prior that would 'naturally' be assumed by the Harvard Medical School Students, upon whom much implicit scorn has been poured [6, pp. 52–54]. The fame of this particular case is based on the fact that a (small) group of students at Harvard Medical School allegedly systematically got the 'wrong' answer when asked what the probability is that x

But how should we interpret the probabilities involved in the above probabilistic reconstructions of 'the' NMA—in particular (a) the probability that evidence e would not occur if theory T were false (or not even approximately true), and (b) the 'prior' probability that T is true (or approximately true)? Any attempt to model these probabilities along the lines of those in the diagnostic case would surely be misguided from the start. In order to develop such a model, we would have to think of ourselves as drawing a theory at random from some population of theories and noting whether it was true,[11] how probable it made e and so on.

Notice then that even if we intend to be 'retailist' about the NMA and concentrate on particular successes for particular theories, any attempted probabilistic reconstruction of the argument along these lines forces us back toward at least a somewhat wholesale view: there has to be some reference class of theories, from which the particular theory is regarded as having been drawn and whose characteristics will play an essential role in the argument. But what population of theories, what reference class, should it be?

Despite the intrusion of some wholesale element, it is surely sensible to minimise that element so far as possible. Certainly, then, this reference class of theories should not be thought of as consisting of 'every possible theory' (of what?) —in part because we have no real grasp on what that might be; and also because, in assessing the impact on, say, Fresnel's theory of light of its success with the white spot, there is clearly no interest at all in the fact that theories from, say, chemistry or biology or even other branches of physics fail to entail that same experimental result (why should they?). Moreover, and in line with my criticism of the wholesale approach, neither is there any interest in how many theories from those other scientific fields are true and/or 'successful' in some generic sense.

A more sensible suggestion seems to be that the reference class should consist of rivals to the specific theory for whose likely approximate truth we are arguing. But how liberal should we be with what we count as a rival? It is well known (and strongly emphasised by Howson [6]) that if we count 'gruesome' alternatives, or, in the case of mathematically expressed theories, Jeffreys-style alternatives [12] as rivals, then that class of alternatives will be infinite, indeed non-denumerable. Moreover it

---

has the disease, given that x tested positive (using similar probabilities to those given above). But one assumption involved in the claim that they got the answer about the posterior 'wrong' is that the 'true' base rate that they 'ignored' is the population incidence of the disease. However, no clinician would intuitively 'model' the event of someone's coming through her clinic door as representing a random selection from the population. People don't attend clinics for no reason—the very fact that they are there means that the reasonable guess about the pre-test probability that they have some disease relevant to the clinician's speciality is considerably higher than the population prior. Even in US medicine, where over-testing is rife, the appropriate prior that a patient has some disease ahead of her being subjected to some test, is—thankfully—seldom, if ever, the overall population prior. (For an antidote to the over-investigation venom see [5].)

[11] Of course truth is not an effective notion and so there are bound to be difficulties here too.

[12] Suppose our theory T links two variables and is of the simple form $y = f(x)$; it predicts that when x takes the value x_0, y will take the value y_0; while when $x = x_1$, $y = y_1$; these predictions turn out to be correct when observations are made; Jeffreys pointed out that there are indefinitely many alternatives T' which share this predictive success (at least in the sense that they equally well

is equally well known that major, surely in fact insuperable, difficulties face any attempt to argue that there is an objectively correct prior probability that a theory drawn from such a set of alternatives has some particular property—say (approximate) truth. As for the other crucial probability in probabilistic formulations of the NMA, namely $p(e/\neg T)$ (or, still worse, $p(e/\neg A(T))$), we might start to think of it as measured by the ratio of all possible alternatives to T (or, still murkier, all possible alternatives to $A(T)$) in which e holds compared to all such alternatives. But aside from the fact that we again have no real grasp on what the set of alternatives is, the standard Laplacian chances approach here—as Colin Howson points out [6, p. 46] —is crucially dependent on the assumption that all the basic alternatives are of equal initial weight, and that is surely preposterous in this case.

So in order to arrive at a sensible probabilistic construal of the argument, we would need to restrict in some way the class of rivals to T (or to $A(T)$) that count as part of the appropriate reference class. But how exactly and with what justification? If we restrict the class of alternatives to T's active rivals at the time of its predictive success, this will normally consist of just one theory T' (the corpuscular as opposed to the wave theory of light, classical as opposed to relativistic physics, etc) and $p(e/\neg T)$ is then identified as $p(e/T')$. In the most straightforward case, where we take the theory T' to come along with all the relevant (currently) accepted auxiliaries, then T' will standardly deductively entail $\neg e$. Thus the corpuscular theory of light with natural auxiliaries entails that there will be no 'white spot', classical physics, again with natural auxiliaries, entails an incorrect motion of Mercury's perihelion, which is however correctly accounted for by relativity theory etc. It is easy then to show that the probabilistic version of the NMA goes through without fallacy, since $p(T/e) = 1$. The argument just becomes the probabilistic version of the deductive rule of disjunctive syllogism (and corresponds in the diagnostic case to there being no false positives, which of course means that any person who in fact tests positive must have the disease, *irrespective of base rates*).

But the term $p(e/\neg T)$ in the probabilistic versions of the NMA cannot in fact simply be identified with $p(e/T')$ where T' is T's main historical rival (if, that is, the reconstruction is to capture the underlying intuitions). The possibility that haunts all versions of the NMA is not that some already available theory, different from T, might share the predictive success e at issue—this will demonstrably not be the case.[13] Instead the worry is that some other, *so far unarticulated*, theory could also predict e, while being radically different from T. No one would claim that it was a 'miracle' that T would get some prediction right if it were false, in cases where some known rival T' (that is, a theory that entails that T is indeed false) also made the same prediction. But suppose that T's success is unique—no other *available* theory shares that success. The worry for the realist is arguably that T's

entail the data points (x_0, y_0) and (x_1, y_1)): just take T' as $y = f(x) + (x - x_0)(x - x_1)g(x)$ for any non-zero function $g(x)$. (For more details see [6, pp. 40–44].)

[13] This of course presupposes that the alternative is taken with its 'natural' auxiliaries; the whole basis of the Duhem problem is that the rival can always be made to entail e if we are allowed to add to it any auxiliaries that we like.

success only *seems* 'otherwise miraculous' to us precisely because we are unaware of some so-far unarticulated possibility T' that equally well enjoys that predictive success, perhaps has other epistemic virtues, and yet entails that T is way off-beam in terms of what it says is going on at the 'noumenal' level. The fact that this so far unknown T' achieves these feats would—however things may seem to us—entail that it would objectively be no miracle for T to have this predictive success despite being false.[14] Or at least T's success would fail to be 'miraculous' in any sense that should incline us to think it likely to be true. The 'explanation' in this case would presumably just be that T happens to have the same consequence in respect of e as does the—let's suppose—true theory T', despite the fact that T is (we are supposing radically) false as revealed by its clash with the true T'. This would be another kind of 'miracle' if you like, but one entirely compatible with (indeed one predicated on) T's falsity.

It seems, then, that if we try think of probabilities like $p(e/\neg T)$ as expressing the ratio of possible alternatives to T in which e holds to all such possible alternatives, then we get into trouble because we have no real handle on that class and certainly no reason to think that all possible alternatives have initial equal weight; but if we restrict the possible alternatives to those we know about (which we might plausibly think about as roughly equal in initial weight), then we also get into trouble since we get trivial answers that have nothing to do with the real issues addressed by the NMA.[15]

4.3.3 The Correct Way to Think About the NMA: The Importance of Not Expecting Too Much

The only serious conclusion to be drawn from the preceding sub-section, so it seems to me, is that there is no available formal probabilistic reconstruction of the NMA that is in anyway convincing because there never was any prospect of producing such a reconstruction. The other proposed reconstruction of 'retail' applications of the NMA to particular successful scientific theories involves interpreting them as 'inferences to the best explanation'. As will perhaps already be clear, I cannot see this as adding anything (except perhaps some confusion) to the intuitions. There *is no method* of inference to the best explanation in any recognisable sense of the word 'method'. Instead scientists develop theories in various ways, some of these turn out

---

[14] This is 'the problem of unconceived alternatives' mentioned by van Fraassen and given centre stage in a recent book by Kyle Stanford [19].

[15] The situation is clearly not likely to be improved by resort to some intermediate position concerning the relevant 'population' of theories—as do Magnus and Callender [14] in identifying this with the class of 'all candidate theories'. Again this set is ill-defined; again it is hardly likely that each candidate theory will sensibly be modelled as carrying the same weight (or plausibility); and again why should the ratio of successful 'candidate' theories that are true (as if we could ascertain this!) in distant fields such as biology or physiology, say, be at all relevant when assessing the impact of the white spot success on the realist credentials of Fresnel's theory?

to be strikingly successful and are 'accepted' as the best available, best empirically supported theories. The suggestion that we are entitled to infer the approximate truth of those theories since it would otherwise be very implausible that they could have been as successful as they have been just is the 'No Miracle intuition'—to think of this as a case of 'inference to the best explanation' adds precisely nothing to that intuition.

Of course it is possible that some other reconstruction can be developed, but it is difficult to see from where. Suppose that the realist in fact concedes that all that she has is the intuition and that she sees no way of a producing a convincing formalisation of the intuitive 'argument' linking striking predictive success to truth. This is indeed a 'concession' that, again following Poincaré, I was always ready to make.[16] It clearly means that the support for scientific realism (and hence for the structural version I advocate) is modest. But is it entirely non-existent?

That conclusion should be resisted. We should, it seems to me, not expect too much from arguments in philosophy, especially at such a fundamental level as this. There is, of course, no question of a theory's predictive success—no matter how startling and impressive—*proving* that that theory is true (or even 'approximately true') and hence solving the problem of ('vertical') induction (or 'abduction', if you like) at stroke! Perhaps William Whewell believed so. He claimed that the predictive successes enjoyed by the wave theory of light were 'beyond the power of falsity to counterfeit'. But of course they are not *provably* beyond the power of falsity to counterfeit: the truth *may* be something radically different from what any current theory says it is, and it goes without saying that the (complete) true theory will have all the right empirical consequences, including those describing the predicted effect at issue.[17]

Can we expect to show that, although it is of course possible that the truth is very different from what our current theories say it is, this is at least extremely improbable in the light of their predictive success? Well again surely not in any objective sense of probability—the process of theory production and evaluation, as I have argued, just cannot plausibly be modelled as involving the drawing of theories at random from some super-urn of 'all possible theories', or even of all possible rivals to some given theory. We have seen why in some detail in the previous subsection, but I think it ought, on reflection, to have gone without saying.

Proofs and objective probabilities are not what 'the NMA' is about. The impact of predictive success, together with the notion of 'approximate truth', is ineliminably intuitive—it is of course *possible* that our current theories are radically false despite their predictive success, but this seems so downright *implausible*. Implausible enough to set realism as the default position. It is surely on reflection

[16] In my [21] I refer to the No Miracles 'consideration', allowing that it is a mistake to regard it as much of an argument.

[17] Though even Whewell can, I think, plausibly be interpreted as holding only that this is not a 'realistic' (as opposed to a merely logical) possibility. Of course we know he was wrong since both Maxwell's theory and photon theory also enjoy the successes at issue and both entail that the classical wave theory is false. But that takes us into the realm of the pessimistic induction.

not surprising that the implausibility here cannot be captured by any sensible analysis in terms of objective probabilities. Realists may wish for something stronger from 'the' NMA , but nothing stronger is defensible. It is salutary here to remember Poincaré's surely correct claim that the NMA intuition is involved, not just in the argument for realism about our scientific theories, but also implicitly in 'ordinary inductive generalisation'. We have learned to expect that there is no solution of 'the problem of induction' (in the original Humean form) either in the form of a convincing deductive argument (by the definition of deductive validity this is *bound* to prejudge the issue!) or in the form of a correct probabilistic argument leading to the conclusion that the generalisation at issue is objectively highly probable, given all the instances. Nonetheless we do not doubt that the reasonable, default, view is that the observational generalisations sanctioned in mature science are in fact correct (though notice that, as Poincaré's case of Kepler exemplifies, this 'correctness' needs to allow for the generalisation's turning out to be strictly false but still 'correct within certain limits'). Similarly in the case of the acceptance of 'observation- transcendent' theories, which Poincaré—again surely rightly— regarded as simply part and parcel of the same process: the fact that we have no proof and no argument for high objective probability does not imply that, again in appropriate circumstances, the reasonable default position is anything other than that those theories are at any rate approximately correct (a position which also allows—as of course does structural realism—for those theories to turn out to be strictly false but still 'correct within certain limits').

So I want to claim that the No Miracles intuition does no more, though also no less, than set some sort of realism as the default position and that it needs no more formal representation in order to do so. Like all arguments for 'default positions', the 'argument' from some theory T's predictive success to its approximate truth is defeasible. And indeed it would clearly be defeated *either* by a demonstration that rival theories sharing T's predictive success but entailing that T is 'radically' false can readily and automatically be created; *or* by the demonstration that there are indeed lots of theories from the history of science that were genuinely predictively successful but which can, by no stretch of the imagination, still be seen as 'approximately true'.

Is there, as some have argued, a demonstration of the automatic availability of 'equally good' rivals to accepted theories? Well, as noted earlier, there certainly are well-known constructions that provide alternatives to any given observational generalisation (grue-style constructions) or to any given mathematically expressed theory (Jeffreys-style constructions)—alternatives that share the same empirical consequences as are taken to support the initially given generalisation or theory. But is the fact that these alternatives, by construction, share the same established empirical consequences as their originals enough to establish that they are 'equally as good' as those originals? Notice that Poincaré, in the passage quoted from Kepler, talks of its being an unacceptably remote coincidence if all of Tycho's observations had the planets agreeing with Kepler's *simple* law and yet—just when neither Tycho nor anyone else was looking— they deviate from their elliptical paths. ('[I]f a *simple* law has been observed in several particular cases, we may legitimately suppose that it will be true in analogous cases. To refuse to admit this would be to attribute an inadmissible role to chance.') The gruefied or Jeffreys-style theories by construction are all ad hoc and hence do not have the simplicity (or more properly unity) that Poincaré required before they are taken seriously. On my account of confirmation [23], although they have the same consequences as the originals, the gruefied or Jeffreys-style alternatives, unlike the original, gain no empirical support from the phenomena that those consequences describe. This is because both constructions involve parameters that are fixed entirely on the basis of those phenomena. Hence those constructed theories are not in fact 'equally good' as their originals. Admittedly an intuitive judgement lies hidden in this account. This is especially clear in the 'grue' case, since, as everyone knows, if we take grue and bleen as our primitive predicates, then it is the 'all emeralds are green' hypothesis that requires specification of the time parameter on the basis of the observations. We just do need to take for granted some intuitions about which theories in which languages are simple or unified. But again: this is philosophy, we should not expect any more.

Bayesians might seem to supply more, but the appearance is illusory. Bayesians can of course endorse the judgement that the grueified and Jeffreys-style constructions fail to count as 'equally good' as the originals out of which they are created. They can do this simply by pointing out that this will automatically be so if these constructions have considerably smaller prior probabilities than the originals.[18] Similarly Bayesians can endorse Poincaré's account of induction by translating his claim that Kepler's first law is simple into an attribution of reasonably high prior probability to it. In general, as Colin Howson emphasises, there is no problem in supplying a Bayesian reconstruction of the NMA, once any attempt to 'objectify' the argument has been abandoned (for the reasons rehearsed earlier). The fallacy that Howson and following him Magnus and Callender exhibit in probabilistic reconstructions of the NMA is obviously blocked if, far from ignoring the 'base rate', a further premise is incorporated into the argument: a premise that asserts that the prior probability of the theory concerned is not low, but in fact reasonably high.

But of course this Bayesian analysis neither eliminates nor in any sense explains the intuitive judgments involved *either* in the counter to the grue/Jeffreys constructions *or* in the NMA. This is because the evaluation of the prior probability is, on the personalist Bayesian approach advocated by Howson, simply a reflection of a personal judgement about the plausibility of the theory. This means that the Bayesian account is certainly not an improvement on Poincaré's and indeed it seems to me a step backward. The sort of judgment of simplicity or unity that Poincaré pointed to, while it may well be 'subjective' in the sense that it is a judgment that scientists apply without being able to explicate it in more basic terms (and certainly not in terms of objective probabilities), is nonetheless universal within science. It seems to be part of science's very ground-rules that theories with parameters adjusted ad hoc to fit the facts are dispreferred to theories that yield the same facts without the

[18] See [7, chapter 7].

resort to such adhoccery. It seems then to be a mistake to regard this as a personal judgment which an individual 'agent' is free to endorse or reject as she sees fit.

Re-focussing on the issue of whether the realist default is defeated by the ever present possibility of constructing empirically equivalent rivals: once this sort of unity or non-*ad hocness* (whether or not regarded as underpinning a high Bayesian prior) is required, then any suggestion evaporates that there are automatic guaranteed ways of generating 'equally good' rivals to accepted theories that entail the 'radical' falsity of those accepted theories.[19] The remaining threat to the realist default is then the more down to earth or constructive one based on the history of theory change. The worry is that the realist position is defeated by the existence of a long list of actual theories from the history of science that were predictively successful but that cannot any longer sensibly be regarded as even approximately true. I deal with this issue directly in a separate [26] paper. Notice however that *if* this worry can be laid to rest by showing that there is a genuine, if sophisticated, sense in which, despite the considerations raised by history of theory change, the development of science has in fact been 'essentially' cumulative, then the default set by 'the' NMA becomes stronger. If whenever a theory, despite its predictive success, is eventually replaced, it is invariably replaced by a theory that not only enjoys still further predictive success but substantially retains its predecessor, then the idea that it is very unlikely that our theories fail to be on substantially the right lines surely becomes still more plausible.

This is of course exactly what Structural Realism claims; and it claims that the 'substantial retention' occurs at the level of structure.

4.4 Conclusion

In this paper, I have argued only that, despite being used as the starting point for a number of more precise arguments that should never have been taken seriously, the facts about the startling predictive success of some of our theories and the intuitive judgments they elicit still count for something. They provide the very modest basis for a very modest realism. No one should claim that realism can be established in any sense, but the success of some of our theories still seems to make realism about them the most plausible default position. Whether, despite the difficulties that have been raised, structural realism can continue to be defended as a position that not only fails to make the success of our theories a gigantic coincidence , but also, far from being defeated by the facts about theory-change in science, gains support from them, is the subject of forthcoming papers [25, 26]. This is, *contra* Magnus and Callender, not a question that should fill any philosopher of science with 'ennui'!

---

[19] See also my [24].

References

1. Boyd, R. (1984) The Current Status of the Scientific Realism Debate. In J. Leplin (ed.), *Scientific Realism* (pp. 41–82). Berkeley, CA: University of California Press.
2. Demopoulos, W. and M. Friedman (1985) Critical Notice: Bertrand Russell's The Analysis of Matter: Its Historical Context and Contemporary Interest. *Philosophy of Science* **52**: 621–639.
3. Duhem, P. (1906) *The Aim and Structure of Physical Theory*, P. Wiener (trans.), Princeton, NJ: Princeton University Press, 1954.
4. Fine, A. (1984) The Natural Ontological Attitude. In J. Leplin (ed.), *Scientific Realism* (pp. 83–107). Berkeley, CA: University of California Press.
5. Gigerenzer, G. (2002) *Reckoning with Risk: Learning to Live with Uncertainty*. London: Penguin.
6. Howson, C. (2000) *Hume's Problem*. Oxford: Oxford University Press.
7. Howson, C. and P.M. Urbach (2007) *Scientific Reasoning: The Bayesian Approach*, 3rd edn. Chicago and La Salle, Il: Open Court.
8. Kahneman, D. and A. Tversky (1972) Subjective Probability: A Judgement of Representativeness. *Cognitive Psychology* **3**: 430–454.
9. Ketland, J. (2004) Empirical Adequacy and Ramsification. *British Journal for the Philosophy of Science* **55**: 287–300.
10. Laudan, L. (1981) A Confutation of Convergent Realism. In D. Papineau (ed.), *The Philosophy of Science.* (pp. 139–165) Oxford: Oxford University Press, 1996.
11. Lewis, P. (2001) Why the Pessimistic Induction is a Fallacy. *Synthèse* **12**: 371–380.
12. Lipton, P. (2000) Tracking Track Records. *Proceedings of the Aristotelian Society.* Supplementary Volume **LXXIV**: 179–205.
13. Lipton, P. (2004) *Inference to the Best Explanation*, 2nd edn. London: Routledge.
14. Magnus P. D. and C. Callender (2004) Realist Ennui and the Base Rate Fallacy. *Philosophy of Science* **71**: 320–338.
15. Newman, M.H.A. (1928) Mr. Russell's Causal Theory of Perception. *Mind* **37**: 137–148.
16. Poincaré, H. (1905). *Science and Hypothesis*. Reprinted, New York: Dover, 1952.
17. Psillos, S. (1999) *Scientific Realism—How Science Tracks Truth*. London and New York: Routledge.
18. Putnam, H. (1978). *Meaning and the Moral Sciences*. Boston MA,: Routledge and Kegan Paul.
19. Stanford, P.K. (2006). *Exceeding our Grasp: Science, History and the Problem of Unconceived Alternatives*. New York and Oxford: Oxford University Press.
20. Worrall, J. (1989) Fresnel, Poisson and the White Spot: The Role of Successful Predictions in the Acceptance of Scientific Theories. In D. Gooding, T. Pinch, and S. Shaffer (eds.), *The Uses of Experiment* (pp. 135–157). Cambridge: Cambridge University Press.
21. ———— (1989) Structural Realism: The Best of Both Worlds. Reprinted in D. Papineau (ed.), *The Philosophy of Science* (pp. 139–165). Oxford: Oxford University Press, 1996.
22. ———— (2000) Tracking Track Records: A Response to Lipton. *Proceedings of the Aristotelian Society*. Supplementary Volume **LXXIV**: 206–220.
23. ———— (2002). New Evidence for Old. In P. Gärdenfors et al. (eds.), *In the Scope of Logic, Methodology and Philosophy of Science* (pp. 191–209). Amsterdam: Kluwer.
24. ———— (2011) Underdetermination, Realism and Empirical Equivalence. *Synthèse* **180**(2): 157–172.
25. ———— (forthcoming) Defending Structural Realism, Or: The 'Newman Objection' What Objection?
26. ———— (forthcoming) Avoiding Pessimism the Structural Realist Way.

References

Part II
Structural Realism and the Nature of Objects and Relations

Chapter 5
Underdetermination as a Path to Structural Realism

Katherine Brading and Alexander Skiles

5.1 Introduction

Two general arguments for structural realism have dominated the literature. The first is inspired by John Worrall's claim that the view offers the "best of both worlds" when it comes to the issue of radical theory change [28]. The argument has an *epistemic* conclusion, suggesting a retreat from the full range of realist commitments about what we can know about unobservable entities, to a more modest subset of these commitments, specifically the *structural* features of such entities. Insofar as this argument is successful, it leads to epistemic structural realism (ESR). The second argument considerably ups the ante, moving from an epistemological claim to an *ontological* one. Ontic structural realism (OSR), defended most prominently by James Ladyman and Steven French, is the view that realists ought to endorse the more radical claim that in some sense all there *is* are these structural features. The central argument for this position begins from the so-called "problem of metaphysical underdetermination".[1]

The focus of this paper is the second argument. Originally formulated in the context of quantum physics, the argument has also been applied in the context of spacetime theories, and discussions have typically assumed that the argument is generally sound. Those who have criticized the argument seem to concede that the alleged metaphysical underdetermination *would be* problematic were it to hold, but then go on to argue that it disappears on closer scrutiny of the particular theory in question.[2]

[1] For presentations of this argument, see [9, 11, 13, 21, §1.2].

[2] For example, see [5, pp. 158–160, 7, pp. 30–31, 16, 17]; for criticism of this type of response, see [10]. An exception to the trend is [22].

K. Brading (✉)
Department of Philosophy, University of Notre Dame, Notre Dame, 46556 IN, USA
e-mail: kbrading@nd.edu

E.M. Landry, D.P. Rickles (eds.), *Structural Realism*, The Western Ontario Series in Philosophy of Science 77, DOI 10.1007/978-94-007-2579-9_5,

In this paper we pursue a different line of inquiry: we examine whether, *even if* the alleged underdetermination is granted, we ought to embrace OSR. After outlining what we take the argument from metaphysical underdetermination for OSR to be (Section 5.2), we offer three criticisms, all of which grant for the sake of discussion that our fundamental physical theories are metaphysically underdetermined in the way the proponent of OSR insists (Sections 5.3, 5.4, and 5.5). Influential as the argument has been, we conclude that it is unsatisfactory as a path to OSR (Sections 5.6).

5.2 Stating the Argument

In a well-known passage, Ladyman first stated the argument from metaphysical underdetermination as follows:

> Even if we are able to decide on a canonical formulation of our theory, there is a further problem of metaphysical underdetermination [. . .] In the case of individuality, it has been shown [. . .] that electrons may be interpreted either as individuals or as non-individuals. We need to recognize the failure of our best theories to determine even the most fundamental ontological characteristic of the purported entities they feature. It is an *ersatz* form of realism that recommends belief in the existence of entities that have such an ambiguous metaphysical status. What is required is a shift to a different ontological basis altogether, one for which questions of individuality simply do not arise. [13, pp. 419–420]

In subsequent literature, presentations of the argument have been similarly informal.

We propose to regiment this line of reasoning as a valid argument that proceeds in two stages. The first attempts to establish a *negative* conclusion targeting what French has called "object-oriented" realism [9, p. 168]: roughly, any of the standard realist views according to which our best fundamental physical theories commit us to objects that are at least as fundamental as the physical structures into which those objects figure.[3] The second stage then attempts to derive the *positive* conclusion that OSR is (all other things equal) preferable to its more traditional counterpart.[4] Here is our re-construction of the argument in more detail:

The argument from metaphysical underdetermination

(P1) Object-oriented realists are committed to objects (that are ontologically basic) having determinate *individuality profiles*: (i) there is a fact of the mat-

[3] French takes the view defended by Psillos [18] as representative. Note that by "object-oriented realist" we also have in mind those who take *neither* objects nor the structures within which they are embedded to be ontologically prior; see e.g. [6].

[4] We take no stance on whether this is the *only* way to formulate the argument, nor do we claim that every proponent of OSR would accept each of its premises. Perhaps there is another way to proceed; we doubt, however, that it will diverge much from the argument we will discuss (though see Section 5.5).

ter about *whether* an object is an individual or not, and (ii) if it *is* an individual, there is a fact of the matter about how, precisely, it is individuated.[5]

(P2) If (P1) is the case, then adopting object-oriented realism commits us to expecting that our best theories will accurately describe what these individuality profiles are like.

(P3) But our best theories fail to offer individuality profiles for their purported objects (much less describe precisely what they are): the individuality status of these objects, as given by our best theories, is *metaphysically underdetermined.*

(C1) So object-oriented realism is (probably) false.

(P4) If OSR is true, then our best theories are not infected with metaphysical underdetermination.

(C2) So, all other things equal, OSR is preferable to object-oriented realism.

As we said before, our strategy is to simply concede premise (P3) for sake of argument, and focus on the status of (P1), (P2), and (P4). That said, it is important to be clear about what (P3) says. This premise should be read as being amenable to a myriad of views about the metaphysics of physical objects. (P3) does not preclude, for example, the possibility of theories according to which some types of object are individuals while others are not; for all (P3) says, the individuality profile of the object-oriented realist's ontology may be very complex indeed. What (P3) *does* preclude however, on pain of metaphysical underdetermination, is a metaphysics of physical objects on which there is no fact of the matter about whether and how objects are individuated (any theory that fails to specify an individuality profile for its objects is necessarily *incomplete*). This will be important in what follows (Section 5.3).

One final point before moving on. The appeals to metaphysical underdetermination are sometimes presented with a *semantic* gloss that our formulation lacks. For example, French writes:

> The imposition of permutation symmetry ensures a kind of *referential inscrutability* which is manifested in both horns of the underdetermination: on the particles-as-individuals view we lose the possibility of specifying which label is associated with which individual; on the alternative, we don't have individuals at all. Only in the former case does some form of causal theory of reference get any purchase, but it's tenuous at best. And given that the physics cannot tell us which case corresponds to how the world is, the question naturally arises: if the realist cannot specify to what it is she is referring—veiled individuals or non-individuals—then to what extent can she claim to be referring to objects at all? [9, pp. 175–176, emphasis added]

In other words, the "underdetermination" prevents the standard realist not only from discerning the individuality profile of her ontology, but even from making out her claim to be referring to objects at all! As far as we know, this version of the

---

5 We leave the restriction to *ontologically basic* objects implicit throughout, for consideration later (Section 5.5).

argument has not been addressed at all in the literature; while not denying its possible importance, we shall not discuss it here.

5.3 Object-Oriented Realism and Premise (P1)

According to premise (P1), a commitment of object-oriented realism is that there is a fact of the matter about *whether* objects are individuals or not—and if so, there is a fact of the matter about *how* they are individuated. In this section, we argue that the object-oriented realist has good reason to reject this premise.

5.3.1 Support for (P1)

French and Ladyman support (P1) by appeal to recent history of physics as well as the testimony of physicists themselves. The first claim is that the concept of object inherited from classical physics involves a commitment to those objects *as* individuals. In their [11], they focus on atomism in the late nineteenth and early twentieth centuries, writing:

> How was the content of atomism cashed out? Or, equivalently, how was the "nature" of atoms understood? Briefly and bluntly put, atoms were understood as individuals where the metaphysical nature of this individuality was typically explicated in terms of substance or, more usually in the case of physicists at least, in terms of the particles' spatio-temporal location. Thus, one of the most prominent and ardent defenders of atomism, Boltzmann, incorporated such an understanding of the nature of atoms in terms of their individuality in Axiom I of his mechanics. The content of atomism was thus cashed out explicitly in terms of the metaphysical nature of atoms. [11, pp. 35–36]

The point is that the axioms that underpin classical Maxwell-Boltzmann statistics include individuality in the concept of object, in the sense that a state and its permutation are counted as distinct states.

The second claim they make is that quantum statistics were seen by the physicists of the time as undermining the view that quantum particles are individuals. The development of quantum mechanics included the development of both Fermi-Dirac and Bose-Einstein statistics, in which a state and its permutation are not counted as distinct states. The connections between individuality, permutation invariance, and different statistics can be challenged,[6] but French and Ladyman are making a case based on history, and their point is that, at the time, physicists saw these developments in quantum mechanics as undermining the view that quantum particles are individuals. French and Rickles briefly summarize the history as follows:

> We shall call this view—that quantum particles are, in some sense, not individuals —the Received View. It became fixed in place almost immediately after the development of quantum statistics itself [. . .]. Thus at the famous Solvay Conference of 1927, Langevin

6 Such as in the work of Simon Saunders; see especially [25] and [26].

> noted that quantum particles could apparently no longer be identified as individuals, and in that same year both Born and Heisenberg insisted that quantum statistics implied that the "individuality of the corpuscle is lost" [...]. Some years later, in 1936, Pauli wrote to Heisenberg that he considered this loss of individuality to be "something much more fundamental than the space-time concept" [...]. [12, p. 221]

The implication French seems to draw from these considerations is that once we are committed to an ontology of particles, we are also committed to providing an individuality profile for those particles (i.e. to discussing whether these particles are individuals or non-individuals). This is the support offered by French and Ladyman for (P1).

Other arguments—from metaphysics and from logic—might be invoked by someone wishing to maintain (P1), and we will discuss these considerations in Section 3.3 below.[7] Our focus here is on the argument for OSR that has dominated the literature, where the support for (P1) is the historical evidence. How convinced should we be of (P1) on this basis? Well, it is not clear that this support for (P1) works, even on its own terms, because it is not clear that French and Ladyman's story is the right way to read the historical evidence.

It is worth saying a few more words about this. We do not dispute that the belief that quantum mechanical particles are not individuals was held to represent an important difference between classical physics and quantum physics, by many of the leading physicists of the time. However, French's own work has shown that quantum mechanics can in fact support an interpretation of its particles as individuals: crucially for the argument from underdetermination, the question of whether quantum particles are individuals or non-individuals is underdetermined by the physics. The physicists that French cites, as believing that quantum mechanics leads to the "loss of individuality" of corpuscles, believed this because they believed that the particles that are the subject-matter of quantum mechanics are described by quantum mechanics in such a way that they are determinately non-individuals. What then, might these same physicists have believed about the status of quantum particles, if they had come to believe that quantum mechanics does not determine whether such particles are individuals or non-individuals? Might they not have concluded that the shift implied by quantum mechanics is not from individuals to non-individuals, but from individuals to particles for whom the categories of individual and non-individual do not apply?

Well, so much for speculative history. Regardless of what they would have said, we think that the latter answer—that there can be objects that are not determinately individuals or non-individuals —is a viable response, and one that should be on the table. (We will come to metaphysical and logical reasons why this might be problematic in a minute, but we will develop this a bit further first.) Consistent with this response is an approach to the objects of physics that we term "law-constitutive".

---

[7] Indeed, French's continued endorsement of (P1) is related to his logical investigations (see [8]).

5.3.2 A Law-Constitutive Approach to Objects

We suggest that a law-constitutive approach to the objects that are the subject-matter of a physical theory is viable, and allows for ontologically basic objects that may be individuals, non-individuals, or not determinately either. This last option asserts that ontologically basic objects may lack an individuality profile, contra (P1). Moreover, such objects need not be such that they can be "structurally reconceived".[8] If we are right, then the rejection of (P1) is consistent with object-oriented realism when combined with a law-constitutive approach to the objects of physics.

It is worthwhile, then, saying a little about the "law-constitutive" approach to the objects that are the subject-matter of a physical theory. That is to say, a necessary (and sufficient, in the strong version) condition of something to *be* a physical object is that it satisfy the laws of that physical theory. The view says something stronger than that to be a certain *kind* of object is to satisfy a certain system of laws. That claim is perhaps uncontroversial. The view we are offering makes the far more radical claim that what it is to *be* a physical object *at all* is to satisfy a certain system of physical laws: there is no concept of physical object that can be given prior to the specification of the laws. This is *not* to say that objects ontologically depend upon our *theories* about what those laws are, or even upon the laws themselves. The proposal is simply that physical theory exhausts all there is to say about what it is to be a physical object: no prior, or theory-independent, conditions of objecthood are to be had.[9]

This gives us a sense in which the physical notion of object has some autonomy from (and can be considered apart from) the metaphysical and logical notions (of which more below). It opens up a third notion of object. Brading [1] has recently argued that the historical roots of the law-constitutive approach go back at least to Newton. Since the support for (P1) offered by French and Ladyman appealed to the authority of historical figures, this history is not irrelevant. Newton is explicit in his writings on natural philosophy that he is offering an account of body suitable for—and restricted to—the purposes he has in mind (his mathematical natural philosophy), and that for something to *be* such a body (a physical body) it must move in accordance with the laws.[10] The claim is that Newton proposed a weak version of

[8] And even if they can, further argument is required to establish (C2), as we discuss below.

[9] There is no guarantee that when we work out the details with respect to the specific laws that we find in this, the actual world, we will arrive at an account of physical object that can indeed serve as their subject-matter. Furthermore, there is no guarantee that this strategy will generate one unified kind of physical body: perhaps the bodies that serve as the subject matter of the laws when gravitation is included will turn out not to be identical to those that serve as the subject matter of the laws when electrical phenomena are at issue. Thirdly, there is no guarantee that the law-constitutive approach to physical bodies will deliver individuals. But these issues are all to be distinguished from the philosophical viability of the law-constitutive approach as a possible account of what it is to be a physical body. We say something more about the issue of philosophical viability below.

[10] For further discussion, see [1].

the law-constitutive view, according to which a necessary condition for something to be a physical body is that it satisfy the laws of motion (but other conditions are also required).

For our purposes, the crucial point is this: if we adopt this approach to physical objects then, unless dictated by the laws, the individuality profile required for physical objects by premise (P1) need not, after all, obtain. We have a viable notion of object of which the object-oriented realist may avail herself, thereby rejecting (P1) as giving necessary conditions on objecthood.

For French and Ladyman, a realism that commits us to physical objects, but fails to determine the individuality or otherwise of those objects, is so strange that they reject it in favor of a commitment to "pure structure" as ontologically basic. Our view is that individuality is distinct from object-hood, and that the "metaphysical underdetermination worry" over individuality can be avoided in a less dramatic-sounding manner. By appealing to the law-constitutive account of physical objects, we can pull apart objecthood and individuality in a very natural way: if a theory makes no commitments concerning whether or not the objects it purports to be about are individuals, then we need not conjoin a metaphysics of individuals versus non-individuals to that theory in order to have a physical notion of object for our theory to be about. In such a case, requiring that we discuss these objects in terms of individuality (and perhaps even commit ourselves one way or another on the matter) demands that we go beyond the content of the theory: we have to add an interpretational layer not warranted by what the theory itself says. Expressed in this way, the alleged "strangeness" of a commitment to objects that is not accompanied by a metaphysics of individuality doesn't sound strange at all—at least not to us.

These grounds for the rejection of (P1) would be cold comfort to the object-oriented realist if the law-constitutive approach necessarily led to structuralism via a different route. French and Ladyman suggest the possibility of a law-constitutive view when writing about Cassirer's structuralism:

> Charge, like the other intrinsic properties, features in the relevant laws of physics and according to Cassirer, what we have here is a reversal of the classical relationship between the concepts of object and law (Cassirer [2], 131-2): instead of beginning with a "definitely determined entity" which possesses certain properties and which then enters into definite relations with other entities, where these relations are expressed as laws of nature, what we now begin with are the laws which express the relations in terms of which the "entities" are constituted. From the structuralist perspective, the entity "constitutes no longer the self-evident starting point but the final goal and end of the considerations [...]". [11, p. 39]

This is a law-constitutive view of the entities that serve as the subject-matter of physics, in a structuralist version, since laws express *solely relations* and objects are *wholly constituted by* these relations.

But notice: whether the structuralist outcome follows from the law-constitutive approach depends on the nature of physical laws, including whether those laws attribute intrinsic properties to objects. In itself, the law-constitutive approach to physical objects is neutral with respect to structuralism: adopting the approach is consistent with, but does not entail, a structuralist reading of the objects that are the subject-matter of those laws.

In sum, the suggestion is that we have a viable concept of physical object that does not entail a substantive further issue about whether those objects are individuals or not. This is a concept of physical objecthood that is consistent with the alleged underdetermination, but which does not necessarily conceptualize objects structurally, and certainly does not eliminate them in favor of structure. It is not our aim here to explicate and defend the law-constitutive approach.[11] Our purpose here is simply to draw attention to this alternative realist position, and to highlight how it can be used to dissolve worries about "metaphysical underdetermination".

5.3.3 *Objects as Individuals: Requirements from Logic and Metaphysics*

The law-constitutive view of objects concerns the concept of physical object, but we have also to address concerns from metaphysics and logic. The questions about what conditions an object has to satisfy have long been given a double-sided treatment, having both a metaphysical and a logical face (think of Aristotle's treatment of individuals). On the metaphysical side, an object must be countable. On the logical side it must be capable of serving as an object of predication.[12]

Do the metaphysical and logical notions require that an object be determinately individual or non-individual, and if so, how does this affect (P1)? The issue of individuality concerns whether an object can be named such that it may be uniquely re-identified at later times and across possible worlds. Our question is therefore whether being countable and/or serving as objects of predication presuppose a fact of the matter about whether the objects in question are individuals.

Consider first the metaphysical requirement that objects must be countable. Traditionally, this rests on the prior condition that objects can be named. However, quantum mechanics casts doubt on this condition. One way to interpret the countability requirement is that there be a determinate answer to the question "how many?". There are numerous examples from quantum theory in which the most natural description of the objects involves numerical distinctness without commitment to naming the objects. Paul Teller [27, p. 128] has a discussion of this issue where he argues for the superiority, in certain contexts, of the Fock space representation of atomic electrons: we model the electrons in a particular atom using occupation numbers, which are numbers describing how many times each property is instantiated, with no regard to "which" particle has which of the properties. In other words, we get the kinds of electron, plus the number of electrons instantiating each kind,

---

[11] That is a much bigger project, to be carried out elsewhere. Among the issues to be addressed are constitution (what precisely is being constituted in any give case (objects, properties, etc.)) and instantiation (the relationship between a theory and its subject-matter, more generally).

[12] If we restrict this to being an object of *sortal* predication the link between the metaphysical and logical aspects, as two sides of the same coin, becomes evident since (on many views) sortal predicates provide conditions for counting.

but no labels enabling us to refer to any one electron in particular—we don't name the electrons. Thus, in quantum mechanics there can be a determinate answer to the question "how many?" independently of whether we name the objects in question.

From this example we see that the metaphysical requirement that objects be countable can be met without appeal to individuality. Thus, the issue of countability is independent of whether the objects concerned are individuals or not. More than that, the requirement of countability does not depend on whether or not the objects are determinately individuals or not.

In order to talk about objects we require a logical notion of object: we must be able to apply predicates. One point in the debate seems to be the claim that the possibility of logical predication depends on appeal to metaphysically robust objects—objects that can be named and then re-identified across possible worlds, and over time. However, Simon Saunders [23, 24] has shown that the logical notion of object, as object of predication, is a weaker notion, requiring only numerical distinctness.

Saunders has argued for a version of Leibniz's Principle of the Identity of Indiscernibles on the basis of which the above example of Fock space poses no problem for the logical notion of object because it admits two-place relations that cope with numerical distinctness for otherwise identical objects: "x is one meter away from y" (for example) gives numerical distinctness by failing to be true when $x = y$ and being true when $x \neq y$. Saunders writes:

> Consider the spherically symmetric singlet state of two indistinguishable fermions. Each has exactly the same mass, charge, and other intrinsic properties, and exactly the same reduced density matrix. Since the spatial part of the state has perfect spherical symmetry, each has exactly the same spatiotemporal properties and relations as well, both in themselves and with respect to everything else. But an irreflexive relation holds between them, so they cannot be identified (namely "...has opposite component of spin to ..."). [24, p. 294]

In Saunders' terminology (following [20]), fermions are "weakly" discernible. Weak discernibility is indeed weak: we cannot refer to one of the two objects in preference to the other. Nevertheless, we can state of the pair that there are two objects, and we can make assertions concerning the properties of those objects. Thus, these objects serve as objects of predication, in the weakened sense given by Saunders' analysis, enabling Saunders to draw the following conclusion:

> I think they [French and Ladyman] are mistaken in their view that failing transcendental individuality, the very notion of object-hood is undermined by particle indistinguishability in quantum mechanics [...] It is true that from exact permutation symmetry it follows that such particles [...] may in certain circumstances not be uniquely identifiable, in the sense that it may not be possible to refer to one member of the collective rather than another. But it does not follow, from logical principles, that such particles cannot be objects of predication. Indeed they can [...]. [23, p. 131]

The logical notion of object does not require that there is a fact of the matter about whether objects are individuals or not. Logic *is* sensitive to individuality and lack thereof: if we enrich our language by adding names to our objects, this will be relevant logically in certain contexts. However, this sensitivity is not relevant to the point that we wish to make here: serving as an object of predication does not depend on prior determination of the status of the object with respect to individuality.

In sum, metaphysics and logic appear to require countability as a condition of objecthood, but not individuality.[13]

The conclusion Ladyman draws from this is that structurally reconceived objects are admissible, but that the scientific realist's objects (since they must satisfy the individuality criteria of (P1)) are not. We claim that this inference is not justified, given the availability of the law-constitutive alternative. We shall say more about this in Section 5.5.

5.3.4 Numerical Diversity as a Criterion of Objecthood

Pooley [17] also argues that the realist should focus on *numerical diversity*. For Pooley, however, numerical diversity is sufficient for individuality. Thus, he rejects our characterization of individuality as concerning whether an object can be named such that it may be uniquely re-identified at later times and across possible worlds. Instead, he distinguishes between "haecceitistic individuality" (for which we reserve the term "individuality") and "non-haecceitistic individuality" (for particulars which satisfy the numerical diversity condition, but neither of the further conditions of having transworld and over-time identity conditions.) We think that he is right to focus attention on numerical diversity, but that his association of individuality with numerical diversity is potentially misleading, and that our taxonomy is preferable, for the following reasons.

Adopting Pooley's terminology, metaphysical underdetermination becomes, in the context of spacetime theory, underdetermination between haecceitistic and non-haecceitistic interpretations of the individuals that are the spacetime points of spacetime theory. This, Pooley suggests, is not a troubling form of underdetermination: the non-haecceitistic interpretation is the "core" realist position, which the realist may or may not supplement with haecceities. He writes: "If this is the only choice to be made, it hardly constitutes an interesting threat to the realist's belief in the existence of spacetime points" [17, p. 10].

A mere re-labelling of both options as concerning "individuals" rather than "individuals" versus "non-individuals" does not make a legitimate underdetermination go away, of course. If such underdetermination exists, then the realist faces a systematic underdetermination of the identity conditions of its basic constituents (the spacetime points); if this seemed troubling on Ladyman's "individuals versus non-individuals" formulation, then it remains troubling on Pooley's "haecceitistic versus non-haecceitistic individuals" formulation. We should make it clear that Pooley never implies that the re-labelling removes the problem; for him, the re-labelling

[13] Bosons, in contrast to fermions, do not satisfy this countability requirement.

is a preliminary step towards denying the alleged metaphysical underdetermination for the theories he considers.[14]

The advantage of our approach, and our terminology, is that we use the term "object" as neutral between (a) haecceitistic individuals, (b) non-haecceitistic individuals, and (c) particulars for whom there is no determinate fact of the matter as to whether they are haecceitistic or non-haecceitistic individuals. This is important because it includes the possibility of objects with no determinate individuality profile. As a result, it allows for the possibility that we do not have to decide between (a) and (b): that we lack the resources to do so does not indicate a metaphysical underdetermination; rather, it indicates that (c) is the appropriate category for the objects of the theory in question.

Pulling together what we have learned so far, the conclusion is that the object-oriented realist has available a route for rejecting (P1), should she so choose. The considerations of this section, including the work of Saunders and Pooley, show that *countability* is the crucial criterion for objecthood. This, and not individuality, is the test for objecthood that can be brought to bear on candidate "objects" arising from the law-constitutive approach to physical objects.

5.4 ESR, Object-Oriented Realism, and (P2)

The conclusion (C1) of the argument from metaphysical underdetermination is the rejection of object-oriented realism. There are two targets here. The explicit target is the standard scientific realist, with her full range of beliefs concerning unobservable objects. However, if the step from (C1) to (C2) is to succeed, the epistemic structural realist option must also be ruled out by the rejection of object-oriented realism. In our opinion, the advocate of ESR can—and should—reject (P2).

First, (P2) is nothing more than the *denial* of ESR, which Ladyman himself casts as a form of object-oriented realism, accepting a world consisting of "intrinsically individuated objects with intrinsic natures" [14, p. 28]. What (P2) says, recall, is that being an object-oriented realist entails that not only that each object has a determinate individuality profile, but also that this profile can be uncovered by the physical theories which describe and refer to them. Yet according to proponents of ESR, this is precisely what the pessimistic meta-induction shows us we cannot have (Section 5.1): according to them, we *cannot know* what the objects these theories refer to and describe are intrinsically like.

---

14 Pooley argues persuasively that his non-haecceitistic version of substantivalism is clearly the better interpretation of current spacetime theory than the haecceitistic one, and therefore that there is no genuine metaphysical underdetermination facing the realist when it comes to the status of spacetime points. Pooley also rejects the alleged underdetermination with respect to quantum particles. He is among those who reject French and Ladyman's claim that a (haecceitistic) individuals interpretation is a genuine option, arguing that quantum particles satisfy numerical diversity but fail both transworld and over-time identity.

But now the problem with the argument's inclusion of (P2) should be clear. On one hand, proponents of ESR deny that we can—much less *should*—expect our best theories to assign an individuality profile to each (type of) object, since part of what the objects of our best theories are intrinsically like is how they are "intrinsically individuated", in Ladyman's phrase. But on the other hand, (P2) just takes for granted that our best theories can provide us with epistemic access to facts about these objects are intrinsically like: in short, the premise just states that ESR is false. For it denies the characteristic epistemological claim of this view, by claiming that we should expect our best theories to describe the intrinsic nature of physical objects. In the absence of considerations supporting (P2), which have yet to be offered, we have a stand-off between ESR and OSR. Insofar as the argument from metaphysical underdetermination is intended to push us beyond ESR, and to OSR, it cannot be the *argument* that is doing the work: (P2) suffices.

Note that this route for rejecting (P2) is open not just to advocates of ESR, but also to *any* object-oriented realist who claims that we have limited epistemic access to the individuality profiles of the objects of physics. Even object-oriented realists who are realist with respect to the intrinsic qualitative properties attributed to objects by our best scientific theories may reject (P2). For instance, consider the form of realism offered by Psillos [19].[15] On this view, though we do have epistemic access to more than the structural content of our theories, we do not have enough to secure knowledge of the world's underlying individuality profile. The dispute is then over whether this epistemic cautiousness amounts to an "*ersatz* form of realism", as Ladyman [13, p. 420] alleges, or to a discovery about our epistemic situation in the world (be that in principle, or contingent on this particular point of our scientific development). In short, the dispute is no longer over whether metaphysical underdetermination poses a difficulty to object-oriented realists. Rather, at issue is whether forms of object-oriented realism that are epistemically humble enough to reject principles like (P2)—by far the most popular strain in the recent literature—are "realist" enough. If there is a genuine dispute to be had here, it has little to do with metaphysical underdetermination.[16]

In sum, the advocate of ESR should most certainly reject (P2), and the standard object-oriented realist should not concede (P2) without a further fight.

---

[15] Psillos offers a version of scientific realism, but explicitly denies that the scientific realist should be committed to the claim that "two" worlds related by the shuffling of bare particulars are in fact two distinct worlds [19, pp. S19–20].

[16] Notice that a scientific realist who endorses the law-constitutive view of objects will insist that these laws fully determine whether there is a determinate fact of the matter as to the individuality or non-individuality of the objects that are the subject matter of the theory in question; thus, in rejecting (P1), she by-passes the force of (P2).

5.5 Reductive vs. Eliminative OSR and (P4)

Our final criticism of the argument from underdetermination starts off with the observation that there are *two* varieties of OSR available—both of which are reflected in French and Ladyman's presentations of the view.

When articulating OSR, French and Ladyman at times appear to argue for the *elimination* of objects in favor of structure. In early presentations, the metaphysical conundrums that OSR attempts to avoid require "a shift to a different ontological basis altogether" [13, p. 420], one in which objects play nothing more than "a heuristic role allowing for the introduction of the structures which then carry the ontological weight" [7, p. 204]. Once they have played this heuristic role, "the objects can be dispensed with" [11, p. 42].

In other presentations, however, there is a shift in terminology towards "reconceptualization" of objects in structural terms, coupled to the suggestion that structures are *ontologically prior* to objects.[17] The reconceptualization of objects in structural terms is the claim that objects are nothing over and above the "nodes" in a web of relations. As such, it weights the objects and relations equally, and does not give ontological priority to either. One role for the argument from metaphysical underdetermination is, we take it, to get us from this "mere reconceptualization" of objects to a claim of ontological priority for structures (that is, to the rejection of object-oriented realism and the adoption of OSR).

Viewed in this role, the argument proceeds as follows: (i) premise (P1) remains as before: objects must have an associated individuality profile if they are ontologically basic; (ii) yet if, however, objects are ontologically derivative upon structures, no such individuality profile is required; (iii) the objects of physics do not satisfy the individuality conditions; therefore (iv) the ontological priority of structure over objects is established.

Before we get to the main objection, note that this subtle shift (from taking the argument from metaphysical underdetermination to support *eliminative* OSR, to instead taking it to support *reductive* OSR) invites further questions when seen in light of the tenability of the law-constitutive view of objects (Section 5.3). In the absence of an argument from "reconceptualization" of objects in terms of structures to the ontological priority of the latter to the former, the option of remaining an object-oriented realist while giving up the requirement that objects must have a determinate individuality profile appears to remain on the table. However, in their recent treatment of the argument from underdetermination, Ladyman and Ross [15, p. 138] retreat from (P1), endorse reconceptualization, and then comment that this is a "thoroughly structuralist" position. This seems to imply that were the object-oriented realist to concede reconceptualization, she would thereby have conceded the debate to OSR. The option of remaining an object-oriented realist while giving up the requirement that objects must have a determinate individuality profile is, on this account, a delusion.

---

[17] For just one example, see [11, p. 37].

There are two ways to think about this, so far as we can see. On the one hand, if reconceptualization yields ontological parity of relations and relata, then further arguments are required to show that either object-oriented language or relations-oriented language more appropriately "reflects basic ontology", and the object-oriented realist need not concede yet.[18] On the other hand, the advocate of OSR could be claiming that the lesson of reconceptualization is that structure is *ontologically prior* to its decomposition into relations and relata, thereby resisting the push towards the relations versus relata debate. This requires that we distinguish *structuralism* from a commitment to the ontological priority of *relations*, something that has not been the case (at least not clearly so) in the structural realism debate to date.

But set these issues aside: a worse problem with the shift from eliminating objects to "reconceptualizing" them is that it reveals how OSR is a victim of its own argument. Premise (P4), recall, says that OSR is not affected by the metaphysical underdetermination that would (allegedly) infect our best theories if object-oriented realism were true; if it *were*, then adopting OSR would be medicine as lethal as the disease. Now, we have distinguished between proponents of OSR who *eliminate* physical objects from those who merely accord them a *less fundamental* status. But this potential disagreement among proponents of OSR is clearly *no less metaphysical* than the dispute between object-oriented realists about whether objects are individuals or not. For the dispute between eliminative and reductive OSR is a dispute about *what exists*, and these are of course paradigmatically metaphysical disputes, no more settled by the details of the relevant physics than for the object-oriented realists.

Thus by trading in object-oriented realism for OSR, we have traded one pair of metaphysically underdetermined interpretations for another. In short, the claim that OSR does not infect our fundamental physical theories with metaphysical underdetermination—i.e., premise (P4)—is false. Moreover, we can also run the argument from metaphysical underdetermination against (OSR) as follows:

(P5) If OSR is true, then there is a fact of the matter about whether objects exist or not.
(P6) If (P5) is true, then we should expect our best theories to say whether objects exist or not.
(P7) But our best theories fail to say whether objects exist; whether they do or not is underdetermined by the interpretations offered by eliminative and reductive OSR.
(C3) So OSR is (probably) false.

Juha Saatsi [22] has also recently suggested that OSR is a victim of its own argument. He claims that "the structuralist proposal only makes matters worse, for

[18] See the further arguments in [15] for the structuralist position, and the arguments of Chakravartty [5] in favor of object-oriented realism.

with such an alternative structuralist ontology available there would be three instead of two to choose from!" [p. 12]. If this is the case, then premise (P4) must be rejected, and the argument fails to progress to the positive conclusion (C2). Such an outcome might be resisted, as Saatsi notes, if it can be argued that the ability of OSR to accommodate the "common core" of the competing interpretations breaks the impasse. However, we think that the escape from underdetermination does not last long. We reject (P4) because of the metaphysical underdetermination *internal* to the radical structuralist program itself, not simply between it and the object-oriented options. Our best fundamental theories underdetermine the correct metaphysics of structure (elimination versus reduction), whether it can accommodate the interpretations offered by traditional realists or not.

Another response to our objection is that there is considerable dialectical pressure to keep objects around, and this mitigates the ontic structural realist's problem with metaphysical underdetermination. For example, its proponents need a response to this notorious objection: structures are sets of relations, and relations require objects in order to be instantiated; so their ontology, consisting of nothing but relations, is either Platonic or incoherent. However, this objection is no problem for proponents of non-eliminative OSR, for they can claim that structural relations *do* have relata while their eliminative rivals cannot. The slogan "no relations without relata" requires only that there *be* nodes in the structure (which the non-eliminativist accepts), not that the nodes be ontologically independent of the structure (which she denies).[19] Similarly, French reintroduces non-structural, non-fundamental elements into the ontology of OSR in order to deal with various other challenges [9, pp. 178–184]. But again, these considerations for preferring reductive OSR to its more extreme counterpart is certainly not constrained by physics anymore than principles and arguments about the metaphysics of individuality are for object-oriented realists. Why are proponents of OSR allowed to break the metaphysical underdetermination by doing metaphysics, but not realists who are friendly to objects?

In this paper we have taken no stance on whether it *would* be problematic if our fundamental physical theories were metaphysically underdetermined. However, what we have shown in this section is that if it *is*, then OSR offers no escape from it.

5.6 Conclusion

Let us recap. We have presented the argument from metaphysical underdetermination as consisting of three premises (P1–3), a negative conclusion that rejects object-oriented realism (C1), a further premise (P4), and a positive conclusion that asserts OSR (C4). We have argued that the object-oriented realist should reject (P1), and that there is available an account of objects—what we have called the law-constitutive approach—that allows her to do so. We have argued that the epistemic

[19] Chakravartty's own response on behalf of eliminative (OSR) is to claim that rejecting the slogan as conceptually incoherent is "question begging" [3, p. 872].

structural realist should certainly reject (P2), even if she does not reject (P1), and that object-oriented realists of any stripe have reasons to question (P2). Thus, insofar as the argument is intended to remove the competitors of OSR (standard scientific realism and ESR) from the game, it fails. Furthermore, we have also argued that the further premise (P4) needed to move to the positive conclusion, asserting OSR, cannot be maintained.

So what path should the proponent of OSR take from here? One option is to modify our reconstruction of the argument; but the onus is on her to offer an explicit and valid argument in support of her desired conclusions, and to show that her premises do not face the same challenges as we have presented here. We think that the prospects are not promising.

In the face of this negative evaluation, it is important not to throw the baby out with the bath water, and to retain the important insights gained from the OSR program. With this conclusion in mind, it is worth noting that much of Ladyman et al.'s [15] recent structuralist manifesto is independent of the success (or failure) of the argument from metaphysical underdetermination. Nevertheless, this argument *is* the central argument by which OSR was introduced, it continues to play a central role in the discussion, and thus deserves independent scrutiny. Our conclusion is that we should look elsewhere than the argument from metaphysical underdetermination when seeking reasons to adopt OSR.

Acknowledgments This paper was produced as part of a project funded by the NSF (grant number SES-0724383 Brading 201201); we are very grateful for this support. We would like to thank all those present during the "Structuralism in philosophy of science" graduate seminar at Notre Dame, fall semester 2008, and acknowledge the various influences that these seminar discussions had on this paper. We are grateful to Anjan Chakravartty, Scott Hagaman, Stephan Leuenberger, and Juha Saatsi for their detailed comments, and to all those who offered questions, comments, and suggestions when material from this paper was presented at Bristol, Duke, Minneapolis, Oxford, Stanford and Wheaton. Finally, we would like to thank the editors of this volume for organizing the Banff structuralism workshop (2005) and for their efforts in putting this volume together.

References

1. Brading, K. (2011) Newton's Law-Constitutive Approach to Bodies: A Response to Descartes. In A. Janiak and E. Schliesser (eds.), *Interpreting Newton: Critical Essays*. Cambridge: Cambridge University Press.
2. Cassirer, E. (1956) *Determinism and Indeterminism in Modern Physics*. New Haven, CT: Yale University Press.
3. ——— (2003) The Structuralist Conception of Objects. *Philosophy of Science* **70**: 867–878.
4. ——— (2004) Structuralism as a Form of Scientific Realism. *International Studies in the Philosophy of Science* **18**: 151–171.
5. ——— (2007) *Metaphysics for Scientific Realism: Knowing the Unobservable*. Cambridge: Cambridge University Press.
6. Esfeld, M. and V. Lam (2008) Moderate Structural Realism About Space-Time. *Synthèse* **160**: 27–46.
7. French, S. (1999) Models and Mathematics in Physics. In J. Butterfield and C. Pagonis (eds.), *From Physics to Philosophy* (pp. 187–207). Cambridge: Cambridge University Press.

8. French, S. and D. Krause (2006) *Identity in Physics: A Historical, Philosophical, and Formal Analysis*. Oxford: Oxford University Press.
9. ———— (2006) Structure as a Weapon of the Realist. *Proceedings of the Aristotelian Society* **106**: 1–19.
10. ———— (2011) Metaphysical Underdetermination: Why Worry? *Synthèse* **180**(2): 205–221.
11. ———— and J. Ladyman (2003) Remodelling Structural Realism: Quantum Physics and the Metaphysics of Structure. *Synthèse* **136**: 31–56.
12. ———— and D. Rickles (2003) Understanding Permutation Symmetry. In K. Brading and E. Castellani (eds.), *Symmetries in Physics: Philosophical Reflections* (pp. 212–238). Cambridge: Cambridge University Press.
13. Ladyman, J. (1998) What is Structural Realism? *Studies in the History and Philosophy of Science* **29**: 409–424.
14. ———— (2007) On the Identity and Diversity of Objects in a Structure. *Proceedings of the Aristotelian Society* **81**: 23–43.
15. ————, D. Ross, D. Spurrett, and J. Collier (2007) *Every Thing Must Go: Metaphysics Naturalized*. Oxford: Oxford University Press.
16. Morganti, M. (2004) On the Preferability of Epistemic Structural Realism. *Synthèse* **142**: 81–107.
17. Pooley, O. (2006) Points, Particles, and Structural Realism. In S. French, D. Rickles, and J. Saatsi (eds.), *The Structural Foundations of Quantum Gravity* (pp. 83–120). Oxford: Oxford University Press.
18. ———— (1999) *Scientific Realism: How Science Tracks Truth*. London: Routledge.
19. ———— (2001) Is Structural Realism Possible? *Philosophy of Science (Supplement)* **68**: S13–S24.
20. Quine, W.V.O. (1976) Grades of Discriminability. *Journal of Philosophy* **73**: 113–116.
21. Rickles, D. and S. French (2006) Quantum Gravity Meets Structuralism: Interweaving Relations in the Foundations of Physics. In S. French, D. Rickles, and J. Saatsi (eds.), *The Structural Foundations of Quantum Gravity* (pp. 1–39). Oxford: Oxford University Press.
22. ———— (2010) Whence *Ontic* Structural Realism? In M. Suárez, M. Dorato, and M. Rèdei (eds.), *EPSA Epistemology and Methodology of Science: Launch of the European Philosophy of Science Association* (pp. 255–265). Dordrecht: Springer.
23. Saunders, S. (2003) Physics and Leibniz's Principles. In K. Brading, and E. Castellani (eds.), *Symmetries in Physics: Philosophical Reflections* (pp. 289–308). Cambridge: Cambridge University Press.
24. ———— (2003) Structural Realism, Again. *Synthèse* **136**: 127–133.
25. ———— (2006) Are Quantum Particles Objects? *Analysis* **66**: 52–63.
26. ———— (2006) On the Explanation for Quantum Statistics. *Studies in the History and Philosophy of Modern Physics* **37**: 192–211.
27. Teller, P. (1995) *An Interpretative Introduction to Quantum Field Theory*. Princeton, NJ: Princeton University Press.
28. Worrall, J. (1989) Structural Realism: The Best of Both Worlds? *Dialectica* **43**: 99–124.

Chapter 6
Kinds of Objects and Varieties of Properties

Antigone M. Nounou

6.1 Preamble

The contemporary debate on scientific structuralism has brought to the fore a major issue: given the continuity of certain structural characteristics, which endure despite theory change, and in light of the theories of modern physics, which lay emphasis on mathematical structure more and more, the metaphysics of physics has to pay structure its due. Views concerning the standing and role of both structure and objects, however, diverge in the various strands of structuralism, and even within each strand.

Among the various structuralists, epistemic structural realists contend that structure is all there is to know; they do not contest the ontological/metaphysical status of objects but insist that their intrinsic natures are unknowable. From the other end of the spectrum, ontic structural realists (OSRs) assert that structure is all there really is; individual objects are relics of a bygone metaphysics. Methodological structuralists, following a middle road, claim that structure and objects meet half way between abstract theory and tangible world. Finally, realists proper insist that structure is determined by objects whose, real, causal properties are non-structural.

Ontic structural realism (OSR) is the most counterintuitive and provocative of all structural realist positions, but mere ascription of adjectives does not constitute helpful criticism. The position as a whole, though, may be criticized for being poorly articulated, as there is no explicit enunciation of what structure is or of what objecthood consists in; and the conclusion for being inadequately supported, since there is little in the literature explaining why exactly the objects posited by contemporary physical theories can and should be re-conceptualized or altogether eliminated from one's metaphysics.

Turning to the explication of 'structure', in the beginning stages of the present-day debate, it appeared as though French [7, 9], advocating OSR, favoured

A.M. Nounou (✉)
National Hellenic Research Foundation, Athens, Greece
e-mail: amnounou@eie.gr

E.M. Landry, D.P. Rickles (eds.), *Structural Realism*, The Western Ontario Series in Philosophy of Science 77, DOI 10.1007/978-94-007-2579-9_6,

set-theoretic structure. Subsequently, however, OSRs further clarified their views on what constitutes the right structure. Thus, they highlighted the importance of group theory and group structure in the structural presentation of fundamental objects of physics, since the latter are presented at the level of scientific practice via group representations (see, for example, [7, 10, 16]). By itself, group structure is extremely abstract, and it can account for nothing physical. But when supplemented with the structure of quantum mechanics (QM), as exemplified by its Hamiltonian or Lagrangian formalism for instance, the structure of quantum field theories (QFTs), and the structure of some spatiotemporal theory, there seems to be enough group theoretically represented physical content in the structuralist construal of theoretical 'structure' for the dialogue to continue.[1]

Assuming that the question 'Which structure?' has been settled in favour of group structure (supplemented with whatever else is necessary), however, does not entail the conclusion 'structure is all there really is', because it is still unclear whether structure thus understood allows for object elimination, or at least for object re-conceptualization in terms of structure alone; and, if it does, whether objects should be so eliminated or re-conceptualized. In order to answer such questions, both OSRs and their critics have to specify which attributes of objecthood play a role in the debate, and then the circumstances under which objects *can* be eliminated or re-conceptualized and those under which they *should* be eliminated or re-conceptualized. The attributes associated with objecthood that recur in the literature time and again include the notions of 'individuation' and 'property'.

In what follows, I first take issue with OSR's insistence that individuation and objecthood are inextricably connected. This presumption has underlain not only the original (and main) argument in favor of OSR but also most of the hitherto rhetoric concerning the relationship between objects and structure. Given that this presumption has been challenged successfully, in my view by Brading and Skiles in this volume (Chapter 5), I rely on their criticism and, in Section 6.2, I use their conclusions to bring to the fore what really matters when OSRs talk about object elimination or reconceptualization: namely, the possibility to eliminate or reconceptualize properties. In a nutshell, my argument in Section 6.2 is as follows. If being an object (physically, metaphysically or logically) does not require being an individual; and if one adopts the conclusion that the fact that a theory may not be able to settle the question whether the physical objects it presents are individuals or not does not entail metaphysical underdetermination, which we'd rather avoid at all costs; then questions regarding the role of, and relations between, structure and objects or kinds thereof ought to be rephrased so that individuation, lack thereof, or even inability to decide about it play no role in the debate.

With an element of confusion thus removed, the fact that properties play *the* decisive role in the debate becomes clear, and so it is to them that our attention turns in the following two sections. The idea that properties come in varieties is

---

[1] French (2009, 15) too admits: "the structuralist still has some work to do in supplementing the [group theoretic] 'object structure' with the relevant dynamical structure".

not new. The distinction between intrinsic and extrinsic properties, relational and non-relational properties etc. have been discussed in the literature quite extensively. A close look at certain aspects of 20th century physics, however, reveals that these characterizations are not particularly helpful to the current discussion as they conceal significant nuances and do not capture what 'structural' means. These issues are discussed in Section 6.3.

In Section 6.4, I rely on QM and QFTs in order to single out three distinct varieties of relational properties and I propose definitions which allow for an appraisal of properties that is pertinent to the current discussion. The appraisal itself, whose importance for the scientific structuralism debate cannot be overstated, remains, however, the topic of another paper.

6.2 The Irrelevance of Individuation

In the years since its first appearance in the literature [14], OSR has split into many distinct versions, which may be thought of as falling into two broad categories: eliminativist and non-eliminativist. Elminativist versions of OSR, asserting that there are no objects, dispense with objects altogether, whereas non-eliminativist versions of OSR, asserting that objects are ontologically secondary, attribute ontological priority to relational structure but maintain that objects should be re-conceptualized as derivative from or secondary to the ontologically primary structure. According to a recent classification [14], the various versions of OSR are associated with the following seven statements[2]:

1. Eliminativism: there are no *individuals* (but there is relational structure).
2. There are relations (or relational facts) that do not supervene on the intrinsic and spatiotemporal properties of their relata.
3. *Individual* objects have no intrinsic natures.
4. There are individual entities but they don't have any irreducible intrinsic properties (incorporates 2 and 3 above).
5. Facts about the *identity* and diversity of objects are ontologically dependent on the relational structures of which they are part.
6. There are no subsistent objects and relational structure is ontologically subsistent (implied by the conjunction of 3 and 4, and also by 5).
7. Constructivism: *Individual* objects are constructs.

[2] To be complete, Ladyman's list should be supplemented with yet another type of OSR proposed by [1], which takes both properties and relations to be ontological primitives but objects to supervene on properties and relations. The proposal, however, is still in embryonic stages and lacks explication as to how exactly one may think of properties as ontological primitives. In my view, there are two options open, none of which seems compelling: either properties attach to some-thing, in which case properties and relations are not the only ontological primitives; or they are free floating, in which case one would opt for one of the already existing metaphysical accounts, like categoricalism or dispositional essentialism, and inherit their problems. For these reasons I chose to leave this account out of the present discussion.

Five of these statements utilize the notions of individuation and identity of objects.[3] This is a reflection of the fact that OSRs have presumed throughout their polemic that object-oriented metaphysics has to take objects to be individuals. Actually, OSR's original (and main) argument which concludes that the outdated metaphysics of *individual* objects should better be replaced by a metaphysics of structure relies very much on this presumably unavoidable relation between objecthood and individuation. This is none other than the (in)famous underdetermination argument, originally put forth by Ladyman [13] and further elaborated by French and Ladyman [10], which, very roughly, goes as follows.[4] Although QM posits objects as part of its ontology, the resources of the theory itself do not suffice to determine whether these objects are individuals nor non-individuals. If objects were part of the ontological store of theories, theories should contain the resources necessary to decide whether their objects are individuals or not. Since our best theories (i.e. QM and QFTs, but also contemporary space-time theories) do not have such resources, we are better off embracing an ontological commitment to structure, a commitment that helps us to avoid the distasteful ontological underdetermination.

In a series of papers, Brading and collaborators ([5] and Chapter 5) target the very presumption that objecthood and individuation are inextricably connected. Brading and Landry [5] argue that when it comes to theories of physics and their ontology, a distinction is needed: although theories of physics *represent particular objects in their models*, they *present kinds of objects* rather than objects simpliciter when taken as a whole. Brading and Collaborators [2–5; Chapter 5, this volume] go further and argue that a viable metaphysical thesis regarding physical objects (the objects that are the subject-matter of physical theories, that is) is that objects are law-constitutive. In their view, this idea goes beyond the view that 'to be a physical object is to satisfy a certain system of physical laws'. Rather, 'The proposal is simply that physical theory exhausts all there is to say about what it is to be a physical object: no prior, or theory-independent, conditions of objecthood are to be had.' Thus their proposal offers a physical notion of objecthood as distinct from metaphysical and logical notions and grants the conclusion that individuation is in fact unnecessary to objecthood, not only from a metaphysical and a logical point of view but also from a physical. For, metaphysically, objects need only be countable, but countability does not require individuation. And from the point of view of logic, all that is required is that objects allow for predication, which, in turn, requires only

[3] In this context, the terms 'identity' and 'individuality' appear to be inseparable, if only because attributing identity to an object presupposes that the object in question can be individuated. Later on, I will use the term 'identity' in reference to kinds of objects, in which case individuation will be irrelevant.

[4] The argument from metaphysical underdetermination has been the subject of various renditions and criticisms. For example, Muller and others, [18–20] disagree with OSRs that QM involves the kind of metaphysical underdetermination posited by OSRs. Ainsworth [1] too argues that the OSR assertion that QM and general relativity underdetermine their object-ontology is problematic, but also that a structural ontology does not necessarily solve the alleged problem. In what follows, we follow the discussion of the argument in Chapter 5, not only because it is more nuanced than the rest but also because it tackles the issue of the relation between individuation and objecthood explicitly.

numerical distinctness that is weaker than individuation. As for the physical notion of object, the implication of their proposal is that if a theory and its laws are silent as to whether the objects they present and hence represent are individuals or not, we too should be silent.

The conclusion of Brading and Skiles has a dramatic effect: not only does it undercut OSR's main motivation, but also it renders the issue of individuation irrelevant.[5] For, it shifts the focus from individual objects to objects unqualified—whether individual or not—or to objects of the relevant kind. As a consequence, the pertinent question becomes whether the metaphysics of physical theories can and, if it can, whether it should be purged not of individual objects, but of objects unqualified or of kinds thereof.

Turning to take the Brading et al. path, we can now investigate how this shift of focus impacts the statements above and hence the OSR versions associated with them. Rather than merely assert that there are no *individual* objects, eliminativists (call them e-OSRs) have to take stock and motivate the elimination of objects yet again. Most importantly, they need to establish that both objects (unqualified) and the kinds of objects postulated by the theories of modern physics can be eliminated, notably by explicating what it takes for objects or kinds thereof to be eliminable. Only then they might silence their critics, who have expressed serious concerns as to whether the very idea of object elimination makes sense at all.

Whether object elimination can be established or not, the point of fact and of interest for the purposes of this paper is that not only elimination but also any other reconceptualization of objects will inevitably implicate the possibility of property elimination or reconceptualization respectively. For, with individuation removed from among the features necessary to objecthood, what is left is properties of objects or kinds thereof.

Consenting to the appropriate modifications, proponents of claims number 3 and 4 above (call them reductionist OSRs or r-OSRs) would now assert that there are particular objects and kinds thereof but that they have no irreducible intrinsic properties. Thus, since the reducibility alluded to is, presumably, reducibility to structure, r-OSRs must demonstrate that the properties of objects or kinds thereof, especially those typically thought of as being intrinsic, are indeed reducible to structure. How exactly this idea might be implemented, we will discuss in the next sections.

Turning to the 5th claim, its advocates should now declare that facts about the diversity of objects and of kinds thereof are ontologically dependent on the relational structure of which they are part. And OSRs embracing the 6th claim would adapt their own assertion to: there are no subsistent objects or kinds thereof but ontologically subsistent relational structure only. Let us call the advocates of both #5 and #6 ontological dependence OSRs or od-OSRs and point out that they must now demonstrate that all facts about the diversity of objects or kinds thereof are ontologically dependent on the relational structure of which they are part.[6] Hence,

---

[5] Note, however, that the issue of discernibility is not rendered irrelevant [8].

[6] Note that in this case it is not only properties but also their specific values that should be ontologically dependent on structure, because it is the specific values that account for the diversity of objects or kinds thereof—but more on this in what follows.

they need to spell out what 'ontological dependence' amounts to and establish that whatever diversifies objects and their kinds ontologically depends on structure.

Finally, constructivists (or c-OSRs) must expound on the meaning of the assertion that kinds of objects are constructs or pragmatic devices.[7]

Interestingly, what was identified at the beginning of this paper as one of the main issues the structuralism debate has raised, namely the need for reassessing the roles of objects and structure in the metaphysics of physics, remains unaffected by this change in vocabulary. For, as OSRs and others have pointed out, QM-entanglement has revealed relations that violate Humean supervenience, and this persists regardless of whether we require objects to be individuals or not. And as we shall see shortly, whatever the objects of physics (fields or particles, individuals or not), certain facts about the identity (not individuation) and diversity of the elementary kinds postulated by QFTs and presumed to exist in the world, appear to be contingent on the existence of other kinds and the specific values of their properties. These facts about the objects of QM and QFTs introduce novelties that challenge long-held presumptions about objecthood. But the resulting need, to reassess the notion of objecthood, does not entail object elimination, or reducibility to structure, or ontological dependence, or construction. For these conclusions, the various OSRs need more. My contention is that OSR supporters of each version must of course explicate what it takes for objects or kinds thereof to be eliminable etc., but the common denominator in all these accounts will be the eminent role that the notion of 'property' has to play.

As I mentioned before, the ensuing discussion draws heavily on QFTs, the theories presenting elementary kinds of objects and representing the (allegedly) elementary kinds of objects populating our world. The ontological picture of QFTs is even less clear than that of QM because, as their name suggests, QFTs involve extended fields instead of localized particles. More work needs to be done in order to understand the idea of objects-as-fields and to determine whether the notion of individuation applies at all in this context and what discernibility amounts to. Even so, having embraced Brading et al.'s conclusion, the issue of individuation has been rendered irrelevant. Furthermore, since even in this context numerical distinctness and countability are not jeopardized, the conversation can continue.[8]

---

[7] To be exhaustive, we should mention that #2 does not constitute a variety of OSR as it only states a fact of QM, which hardly suffices for concluding that structure is all there really is. On the contrary, Morganti [17] argues that we can still talk about individual objects with relational properties meaningfully. As for #3, it mentions natures and I'd rather steer clear off such talk. Moreover, even Ladyman and Ross [15] duly replace talk of natures by talk of properties. Thus, if we replace 'natures' with 'properties' in #3 then we obtain #4, i.e. r-OSR.

[8] The total number of objects is not conserved in QFT processes. This fact, however, does not endanger my claim because the number and diversity of objects in the initial and final states of interacting systems are granted; and so are the number and diversity of objects involved in interactions—'virtual particles' aside. Another possible objection might be that there is no unique vacuum state in QFTs in curved space-time, and that countability fails therein because accelerating observers detect objects that inertial observers don't. A more careful reading of the relevant literature [21], though, reveals that (i) objects manifest in such a vacuum through interactions only;

6.3 Varieties of Properties

To eliminate objects or kinds thereof in favor of structure, or to reduce them to structure, or to show their ontological dependence on structure, or to construct them from structure, one has got to eliminate or reconceptualize accordingly whatever it is that characterizes them. Whether restricting our attention to kinds of elementary objects only or including particular objects too, the pertinent characteristics that need be eliminated etc. are their properties. But what exactly do we mean by 'property'? Do we mean the generic properties, like electric charge for example, that characterize all elementary kinds of objects of QFTs, or do we also have to include the exact values of such properties, like -1, that characterize the specific kinds of, say, the Standard Model and have hitherto been thought to be intrinsic? Do we mean only the properties of the kinds—generic or specific—of objects presented by physical theories, or need we also take into account properties of particular objects, including their spatiotemporal properties which are clearly relational? Do OSRs have to eliminate or reconceptualize some of these properties or all of them? And if the former, how can we separate the wheat from the chaff so to speak? These are non-trivial questions.

Focusing on kinds of objects for a moment, QFTs *present* two generic kinds of elementary objects with mass, leptons and quarks. Leptons are characterized by the properties mass, spin, and electroweak charge, and they are distinguished from quarks by the fact that the latter also carry the property strong charge. Therefore, in order to defend a claim like 'leptons and quarks can be eliminated or reconceptualized' OSRs have to explicate how the generic properties like mass, spin and interaction charges, which characterize each of these two generic kinds, can be altogether eliminated or reconceptualized in terms of structure alone.

Yet again, the Standard Model (SM) of (all but gravitational) fundamental interactions presents leptons and quarks that are subdivided further into specific kinds of leptons and quarks, which *represent* the kinds of elementary objects that inhabit our world. When it comes to these specific kinds of objects, each kind of leptons, such as electrons, is distinguished from the other kinds of leptons (with electric charge), such as muons, only by the fact that the value of the property 'mass' is particular to each kind. Thus, electrons have a mass of 0.5 MeV whereas muons have a mass of 105.7 MeV. Ditto for quarks. Therefore, it seems that in order to defend the claim 'electrons, muons, etc. can be eliminated, or reconceptualized' OSRs have to spell out how to eliminate or reconceptualize not only the generic QFT properties that characterize leptons and quarks, but also their specific values that appear in the SM and distinguish, say, electrons from muons. Since this conclusion might seem a bit rushed, let us reflect on this issue a bit more.

---

hence the theory talks sensibly only about objects that are being detected, and (ii) if a non-inertial observer detects a number of objects, the inertial observer reports the non-inertial observer as having detected the same number of objects. Therefore, numerical distinctness and countability of QFT objects remain intact even in curved space-times.

The difference between the generic properties of the kinds of objects presented by QFTs and the specific properties of the kinds of objects presented by the SM reflects a difference between generic kinds of objects, like leptons and quarks, which are merely presented by a theory and specific kinds of objects, like electrons, muons, etc., which are presented by specific but general models of (or compatible with) the theory. The difference between the two is that the actual kinds of objects populating our world are represented only by the specific kinds of objects presented in specific models like the SM and not by the generic kinds of objects presented in theories like QFTs. For, the theory affords several models like the SM, whose specific kinds of possible objects would represent kinds of objects populating possible worlds that differ from ours in their ontology if not also in some of their laws. The fact that there is such a difference between generic and specific properties entails two distinct responses to the question 'Which properties to eliminate or reconceptualize?' depending on which structure exactly OSRs must commit to. If it is the group structure of QFTs only, then the properties that need be eliminated or reconceptualized are the generic properties only. But then, the structure one is realist about is a generic structure compatible with many possible worlds, including ours. Commitment to the reality of the group structure of the SM, on the other hand, entails realism about the specific structure that purports to represent our world including the kins of objects populating it. The decision may turn out to be arbitrary or conventional, but whatever it is, OSRs will have to spell out how to eliminate or reconceptualize either the generic properties of the kinds of objects of QFTs alone, or (most likely) the specific properties of the kinds of objects of the SM or similar specific models of QFTs if the SM is to be replaced.

Apart from the generic and specific properties characterizing the elementary kinds of objects of QFTs, the objects of QM are characterized by a property that is peculiar to certain QM systems. As QM objects combine and form entangled systems, usually pairs, they are represented by so-called entangled states and share the property of being in an entangled state. Do properties like being in an entangled state have to also be eliminated or reconceptualized in terms of structure in order for the OSR program to succeed, and if yes how? These are additional questions that OSRs need address in their quest for object elimination or reconceptualization.

Finally, turning to a theory of gravity—be it Newtonian or general relativistic—and of gravitational interactions, the property 'weight' comes to mind as yet another property characterizing objects. This property, though, characterizes particular objects, which may but usually don't exemplify elementary kinds. In order to eliminate or reconceptualize objects, would OSRs have to also eliminate or reconceptualize properties like weight? The answer to this question depends, I believe, on whether OSRs are committed to thus eliminating or otherwise reconceptualizing particular objects (of some kind or other), which typically manifest in the concrete models of a theory that are used, in turn, in representing physical systems.

The discussion concerning which properties ought to be eliminated or otherwise reconceptualized in order for OSR to go through as a viable metaphysical position has just begun. The point we have ascertained though is that the issue whether objects or kinds thereof can be altogether eliminated, or reduced to structure, or shown to depend on structure ontologically, or constructed from structure boils down to whether certain varieties of properties, possibly all those mentioned above and more, can be thus eliminated, reduced, etc. In the next section, we will examine in more detail the varieties of properties that OSRs should be concerned with and what it would take for them to be eliminated or reconceptualized.

Before we move on, though, a remark is due regarding the classic varieties of properties presented in the literature, namely intrinsic and non-intrinsic, relational and non-relational, qualitative and non-qualitative, interior and exterior. Although the intuitions that have motivated these distinctions seemed clear at first, offering lucid philosophical accounts has been less than straightforward and so has been finding out the alleged correspondences between the different pairs.[9] More pertinent to the current discussion though is that this classification (or classifications) of properties is of little help. For example, although the original r-OSR thesis cites 'intrinsic properties' explicitly, the distinction 'intrinsic-extrinsic' capture the nuances discussed above. And too, the notion 'being relational' does not encapsulate the idea of being structural in the sense of being reducible to structure or of being ontologically dependent on it.[10] It is for these reasons that in the remainder of this paper we will turn our back to the extant literature on the aforementioned distinctions and look deeper into physics instead in order to shed more light not only on the varieties of properties that appear in physical theories and models, but also on the roles these varieties play in the structural realism debate.

6.4 Relational and Structural

6.4.1 Relational-1, -2, and -3

Properties like weight are reducible to—in the sense of being determined by—other properties and spatiotemporal (or external) relations, whereas properties like mass are not. Properties like being in an entangled state, which cannot be thus determined (or be thus reducible), do not supervene on the properties upon which they depend. Finally, there are contingently relational properties, which depend on internal rather than spatiotemporal relations. These facts indicate a diversification of the meaning of relational property, the first variant being exemplified by the property weight.

---

[9] For recent discussions see [6, 12, 22] and references therein.

[10] From now on I put aside e-OSR and c-OSR and focus on r-OSR and od-OSR only. For one, e-OSR has been criticized severely from many, mostly on the basis that relational structure without relata makes no sense. I believe that the only way to make sense of e-OSR is to reduce it to some version of r-OSR or od-OSR if either of the two proves to be viable. It is for this reason that I will ignore it in what follows. As for c-OSR, I really don't know how to translate the term 'being a construct' unless talking about reduction again. Hence in what follows, I stop worrying about c-OSR also.

6.4.1.1 Relational-1

The weight of a particular object is a relational property because it depends on certain relations the object bears to other objects: it differs according to the gravitational field in which the object is embedded. The gravitational field may be thought of as a property that space-time acquires due to mass distribution, which, in turn, is determined by the masses of objects that are in the vicinity of the object in question. Thus, the weight of an object is determined by the object's own mass, the masses of its neighbours and spatiotemporal relations. Let us call properties like weight *relational-1* and define them as follows:

> A property of a particular object is relational-1 if and only if it depends on spatiotemporal relations between the object in question and other objects as well as on properties of the objects involved.

In the definition above, the dependence of relational-1 properties on other properties and relations may be recast in terms of reducibility, thus catering for r-OSR, or in terms of supervenience, thus catering for od-OSR. That is to say, one could claim that the weight of an object is reducible to the masses of the objects involved and the between them spatiotemporal relations; or that it supervenes on the mass of the object itself, the mass distribution around it and the, inevitably, external relations. In both cases, however, it does not follow automatically that relational-1 properties are also structural. For this assertion to be justified, the properties on which relational-1 properties depend (in this case the object's mass and the mass distribution around it) should be also shown to be either reducible to structure or ontologically dependent on it respectively.

Whether the mass of particular composite objects and the mass distribution in the in particular neighborhoods of the cosmos or even the mass of the universe as a whole can be characterized as structural is an interesting question that falls outside the scope of this paper. The point raised here though remains: relational-1 properties are not self-evidently structural and may be characterized as such only if all the properties on which they depend are characterized as structural too. Before we move on, it is worth noting that relational-1 properties are typically presented in concrete models, which may be used to represent systems in the world.[11]

6.4.1.2 Relational-2

A second variety of relational properties is exemplified by the property 'being in an entangled state', which is dictated by an idiosyncratic type of (non-classical) relation a quantum object bears to another quantum object after the two have interacted and formed a system. Some properties of this system are holistic in the sense that they characterize the state of the system as a whole, regardless of the spatiotemporal

[11] Recall that 'concrete models' differ from 'generic models' as follows. Concrete models are used in representing physical systems, and they are distinct from the specific but generic models of (or compatible with) a theory, which typically represent physically possible systems or physically possible worlds.

separation of its parts. Thus, unlike weight, this property is not attached to any one particular object nor does it depend (in the sense of supervening) on the external relations these objects bear to each other. It does depend, however, on certain properties (e.g. spin, helicity, etc.) that characterize the objects which formed the system in the first place or the kinds thereof.

The exact form of the correlation of objects that constitute systems with properties like 'being in an entangled state' is dictated by the structure of the theory, in this case QM. The formalism of QM presents entangled systems involving two objects, say O_1 and O_2,[12] as sums of states in the tensor product Hilbert space $H_1 \otimes H_2$, H_1 and H_2 being Hilbert spaces whose states present kinds of objects like O_1 and O_2 and represent objects like O_1 and O_2. Thus, having the property of being in an entangled state is a specific variety of relational property that the states of correlated quantum objects may have, and though it depends on certain properties that the objects of their kind possess, it does not supervene on external relations. Let us call this variety of relational properties *relational-2* and define it as follows:

> A property attributed to a system of correlated objects (typically comprising two such objects) is relational-2 if and only if it is dictated by the correlation as a whole, and although it depends on certain properties of the kinds of objects involved in the system, it does not supervene on their spatiotemporal relations.

Relational-2 properties are non-intrinsic so far as particular objects are concerned, and one might also contend that they are structural because they are dictated by the structure of QM and hence by the structural aspects of the world represented by QM. One, however, cannot be too careful. Let us note first that in the definition above 'dependence' means, loosely, no more than that the objects (of the kinds) that can get entangled must have certain properties like spin etc. in order to become entangled in the first place. Put even more loosely, no such properties entails no interaction between the objects in question and therefore no entanglement. Hence whether properties like being in an entangled state are also structural depends on the possibility of characterizing generic properties like spin and helicity as structural. To this question we will return below.[13]

6.4.1.3 Relational-3

Yet another variety of relational properties is exemplified by the mean lifetime of muons (2.2×10^{-6} on their rest frames), which appears, at first glance, to be intrinsic: all muons in our world decay spontaneously into electrons and neutrinos in

---

[12] Strictly speaking, QM does not describe individual objects and but rather systems of objects belonging to kinds 1 and 2 respectively (leaving open the possibility that 1 and 2 may be of the same kind also). My rather loose use of the jargon, however, does not affect the point I discuss here.

[13] Note in passing that the properties spin, helicity, etc. referred to above are not projections, specific values, but the generic properties that characterize the objects of QM and QFTs. For a discussion of failure of Humean supervenience involving projections, I refer the interested reader to [12].

(approximately) the aforementioned time.[14] Additionally, the lifetime of any particular muon is independent of its external relations to any other particular objects, elementary or not, that may be present at its environs on its path to decay. Appearances, however, can be deceptive.

Unlike weight, mean lifetime is not determined by, nor does it supervene on properties of particular objects and their spatiotemporal relations. And unlike being in an entangled state, mean lifetime does not depend on relations between particular objects—whether individuals or not—which are dictated by a theory and are presented in both its general structure and its concrete models. Rather, it is relational because upon closer examination it turns out to depend on whether leptons lighter than muons exist in a world. For this reason, mean lifetime may be thought of as being relational but due to internal relations. Let me explain.

According to the Standard Model of elementary interacting particles, there exist three generations of leptons (and their respective anti-leptons) in the world, which are classified by the irreducible representations of the SU(2) × U(1) group of electroweak interactions. These irreducible representations are arranged in so-called weak isospin doublets:

$$\begin{pmatrix} e^- \\ \nu_e \end{pmatrix}, \begin{pmatrix} \mu^- \\ \nu_\mu \end{pmatrix}, \begin{pmatrix} \tau^- \\ \nu_\tau \end{pmatrix} \tag{6.1}$$

and are typically characterized by the spin, mass, and interaction charges of each of the kinds (massive leptons and their neutrinos) that belong to each generation. The corresponding elementary objects described by all three generations are characterized by the same spin and respective charges but differ in the masses of (at least) the heavier leptons, the electrons, the muons and the tauons. The first generation, comprising electrons and their neutrinos, is the lightest, with electron mass at 0.5 MeV. In the second generation, the mass of muons is at 105.6 MeV and in the third, the mass of tauons is at 1776.8 MeV.

When muons decay, they decay to electrons and neutrino-antineutrino pairs. But the fact that they decay at all is contingent upon the existence of at least one other generation of leptons that is lighter than muons. Since for all we know at present (and in spite of fervent aspirations and attempts) the masses of each kind of elementary objects are not determined by any laws or symmetry structure, a world inhabited by electrons alone is physically possible; and so is a world in which muons are the lightest leptons.[15] Thus, each of these worlds, including ours, constitutes a model consistent with QFTs. Additionally, in the world where muons are

[14] Strictly speaking 'mean lifetime' is a statistical property. Hence the actual lifetime of particular muons is not equal to 2.2×10^{-6}. The claim above, however, is accurate enough for the purposes of the current discussion.

[15] Such worlds are physically possible because the masses of the elementary kinds of objects described by the Standard Model are free parameters. Therefore, if muons *truly* were the only or the lightest leptons in a world, no physical law would necessitate or even permit their decay. Masses and charges are not determined by any tentative extensions of the SM either.

the lightest massive leptons, muons could not decay, because there would be no lighter leptons to decay to. Nonetheless, in our world, or in any physically possible world where both muons and electrons exist and interact electro-weakly, all muons decay to electrons in $2.2 \times 10^{-6}s$ and electrons don't decay at all. In yet another physically possible world, that is in yet another model consistent with QFTs and even with the SM, one where leptons lighter than muons but of mass different from 0.5 MeV exist and interact electro-weakly, muons decay, but with a different mean lifetime whose duration depends on and is determined by the respective mass-difference.

Assuming muons exist in a world, their decay is therefore contingent upon what other kinds of leptons exist in that world. Furthermore, since in worlds where muons decay the span of their mean lifetime is determined by (and supervenes on) the difference between the muon mass and the lighter-than-muon lepton mass, mean lifetime is a relational property.

The fact that there exists a difference between the masses of the various kinds of leptons is expressible succinctly as an ordering relation between their respective irreducible representations (in terms of the respective parameters). Therefore, we may attribute the mass-difference upon which the muon's mean lifetime depends to internal relations between irreducible representations. We may distinguish, then, this variety of relational properties from the aforementioned relational-1 properties and relational-2 properties by dubbing them *relational-3* and define them as follows:

> A property of objects of a particular kind is relational-3 if and only if it depends on relations that the irreducible representation of its kind bears to irreducible representations of other kinds.

Relational-3 properties display structural traits. The pertinent relations between the relevant properties (the mass ordering and the corresponding mass differential in our example) can be thought of as internal relations between irreducible representations. Now, these relations, and hence the properties associated with them, are structural in that they involve an aspect of the group structure of QFTs, namely the irreducible representations. But can the properties associated with them be characterized as structural unqualified? I believe that to answer this question one has to spell out what 'dependence' in the definition above amounts to.

Clearly, whether muons decay at all and, if they do, what their mean lifetime is cannot be determined by the generic ordering relation that holds between the irreducible representations of the SU(2) × U(1) group alone. For, mathematically speaking, there exist infinitely many mass values (and hence irreducible representations) between the electron mass $m_e = 0.5\,\text{MeV}$ and the muon mass $m_\mu = 105.6\,\text{MeV}$; and as many between $m = 0$ and $m_e = 0.5\,\text{MeV}$. On the other hand, one cannot appeal to the structure of the SM and claim that the mass values for muons and electrons and hence muon mean lifetime are determined by the structure of the SM, because the masses of the kinds of objects of the SM (both leptons and quarks) are free parameters which are constrained and determined only by the world's contingencies and not by mathematical possibilities.[16] Hence, muon decay and mean lifetime are not reducible to structure alone, unless the determinate values of the relevant properties are themselves reducible to structure.

Similarly, od-OSRs would require that relational-3 properties can be shown to either supervene on structure or depend on it ontologically. But unless we include in the structure the relevant properties' specific values, supervenience of relational-3 properties on the relevant structure of the irreducible representations of SU(2) $\times$ U(1) fails as the mean lifetime of muons can change with the structure remaining intact. That is to say, a change in the value of the mass of the lightest massive lepton would not necessitate a change in the structure SU(2) $\times$ U(1), not even a change in the ordering relation between the respective irreducible representations representing muons and the lightest leptons. And yet, such a change in the electron mass would result to a change in muons' mean lifetime. Therefore, the supervenience assertion fails, unless the exact values of the relevant properties are structural too. A similar conclusion can be drawn if supervenience is replaced with ontological dependence. Therefore, for relational-3 properties to be also characterized as structural, it is necessary to demonstrate that the properties upon which relational-3 properties depend are structural.

6.4.2 Properties Hitherto Known as Intrinsic

So far I have argued that when restricting our considerations to r-OSR and od-OSR, the question regarding object reconceptualization in terms of structure is whether properties can be—respectively—reduced to structure or shown to depend ontologically on structure. The examples from physics have led us to three distinct definitions for relational properties, which also reveal that there is a difference between being a relational property and being a structural property. The difference hinges on whether the properties upon which the three varieties of relational properties depend can themselves be characterized as structural. In relational-1 properties it was properties like mass but of specific and usually composite objects; in relational-2 properties it was properties like spin but of generic kinds of objects; and in relational-3 properties it was specific properties with determinate values but of the kinds of elementary objects that populate our world. The idea that emerges from the discussion so far is that if both the generic and the specific properties upon which relational-1,

---

[16] Indeed, had a QFT or some other theory existed that would deliver the masses of the existing leptons and quarks theoretically/structurally, the masses of the SM and hence the masses of the kinds of objects populating our world would have been determined theoretically/structurally. But adhering to what Ladyman and Ross [15] have called 'principle of scientific closure' and the associated stipulations (pp. 37–38), we are compelled to include to our discussion bona fide 'specific scientific hypotheses' that either have been investigated or might be investigated in the future. This excludes hypotheses like 'the ultimate theory of fundamental objects is within reach' that that fall with science fiction rather than science proper. Hence, given the present state of scientific knowledge I press on.

-2 and -3 properties depend are themselves structural, in the sense of being reducible to or ontologically dependent on structure, then the properties that depend on them too are reducible to or ontologically dependent on structure and therefore structural.

Leaving aside particular composite objects and relational-1 properties, the characterization of relational-2 and relational-3 properties as structural depends, therefore, on the possibility of thus characterizing not only the generic properties mass, spin, helicity and interaction charges, but also the hitherto thought to be intrinsic specific properties, with their determinate values, that characterize the kinds of elementary objects of our world.

Clearly, the irreducible group representations provide a purely structural tool for classifying the generic properties like mass etc. by way of structure, and there is a whole story to be told—elsewhere—how this tool has become available and to what extend these generic properties are group constituted in the sense of being determined by group structure alone. The point though is that we need two more definitions implicating this structure of groups, one that will enable us to characterize the generic properties of mass, spin and interaction charges as structural, and another that will do the same for the specific properties with the determinate values of the particular kinds. The two definitions I propose are as follows:

(a) The generic properties spin, helicity, mass and interaction charges of generic kinds of elementary objects are structural just in case they are constituted group theoretically.
(b) A property whose exact value characterizes a particular kind of elementary objects is structural if and only if it depends exclusively on the structure of irreducible group representations.

In the second definition, the notion of dependence has been left unspecified, as in the previous definitions. Deliberately. Because the dependence relation may be worked out only after one's OSR commitments have been stated explicitly. Thus, r-OSRs will have to establish that the aforementioned properties are reducible to structure, and the od-OSRs that they depend on structure ontologically. Shedding light on this issue and reaching the conclusion as to whether mass, spin and interaction charges truly are structural, however, fall outside the scope of this paper.

In conclusion, assuming that the list of definitions I have provided exhausts all possibilities of characterizing properties as relational and as structural, my contention is that whatever one's favored flavor of OSR, OSR cannot succeed unless the characteristic properties that differentiate the various kinds of elementary objects can be characterized by the last definition and are thus shown to be structural. Of course a lot will hinge on the exact form the dependence that the notion of structural (b) will employ. Be that as it may, if the characteristic properties turn out to be structural, then showing that relational-1, -2, -3 properties are also structural and that objects or kinds thereof can be deflated metaphysically should be straightforward, whatever the flavor of OSR one favors. Tackling this issue, however, will be dealt with elsewhere.

Acknowledgments The work that resulted in this paper begun in preparation for the Banff Workshop on 'Structure, Objects, and Causality'. Earlier versions of this paper were presented at the 5th Meeting of Pittsburgh Fellows, at the Notre Dame workshop 'Structuralism in Physics' and at the colloquium of the department of History and Philosophy of Science at the University of Athens. I would like to thank the organizers and participants of all these events for extremely beneficial (to me at least) comments and discussions. Special thanks go to Aris Arageorgis, Katherine Brading, Elena Castellani, Ron Giere, Steven French, Elaine Landry, Matteo Morganti, F. A. Muller, Stathis Psillos and Vassilis Sakelariou for helpful comments and constructive criticisms. I would also like to acknowledge my substantial intellectual debt to Serge Rudaz. This work owes a lot to his unparalleled grasp of physics and to his eagerness to share this knowledge with me. At its latest stages, the research leading to this paper was funded by the European Union's Seventh Framework Programme [FP7/2007-2013] under grant agreement n°229825.

References

1. Ainsworth, P.M. (2010) What is Ontic Structural Realism? *Studies in History and Philosophy of Modern Physics* **41**: 50–57.
2. Brading (2012) Newton's Law-constitutive Approach to Bodies: A Response to Descartes. In A. Janiak and E. Schliesser (eds.), *Interpreting Newton: Critical Essays* (pp. 13–32). Cambridge: Cambridge University Press.
3. Brading (2011a) On Composite Systems: Descartes, Newton, and the Law-constitutive Approach. In D. Jalobeanu and P. Anstey (eds.), *Vanishing Matter and the Laws of Nature: Descartes and Beyond* (pp. 130–152). Studies in Seventeenth-Century Philosophy. New York: Routledge.
4. Brading (2011b) Structuralist Approaches to Physics: Objects, Models and Modality. In A. Bokulich and P. Bokulich (eds.), *Scientific Structuralism* (pp. 43–65). Boston Studies in the Philosophy of Science, vol. 281. Dordrecht: Springer.
5. Brading and Landry, E. (2006). Scientific Structuralism: Presentation and Representation. *Philosophy of Science* **73**: 571–581.
6. Francescotti, R. (1999) How to Define Intrinsic Properties. *Noûs* **33**(4): 590–609.
7. French, S. (1999). Models and Mathematics in Physics: The Role of Group Theory. In J. Butterfield and C. Pagonis (eds.), *From Physics to Philosophy* (pp. 187–207). Cambridge: Cambridge University Press.
8. ———— (2011) Metaphysical Underdetermination: Why Worry? *Synthèse* **180**(2): 205–221.
9. ———— (2000) The Reasonable Effectiveness of Mathematics: Partial Structures and the Application of Group Theory to Physics. *Synthèse* **125**: 103–120.
10. ———— and J. Ladyman, (2003) Remodelling Structural Realism: Quantum Physics and the Metaphysics of Structure. *Synthèse* **136**: 31–56.
11. Humberstone, I.L. (1996) Intrinsic/Extrinsic. *Synthèse* **108**: 205–267.
12. Karakostas, V. (2009) Humean Supervenience in the Light of Contemporary Science. *Metaphysica* **10**: 1–26.
13. Ladyman, J. (1998) What is Structural Realism? *Studies in History and Philosophy of Science* **29**: 409–424.
14. Ladyman, J. (2009) Structural Realism. In E.N. Zalta, (ed.), *The Stanford Encyclopaedia of Philosophy*. http://plato.stanford.edu/entries/structural-realism/#OSRGroThe.
15. Ladyman, J. and D. Ross, (with D. Spurrett and J. Collier) (2007) *Every Thing Must Go: Metaphysics Naturalised*. Oxford: Oxford University Press.
16. Landry, E. (2007) Shared Structure Need not Be Shared Set-Structure. *Synthese* **158**(1): 1–17.
17. Morganti, M. (2009) Inherent Properties and Statistics with Individual Particles in Quantum Mechanics. *Studies in the History and Philosophy of Modern Physics* **40**: 223–231.
18. Muller, F.A. (2011) Withering Away, Weakly. *Synthèse* **180**(2): 223–233.

19. Muller, F.A. and S. Saunders, (2008) Discerning Fermions. *British Journal for the Philosophy of Science* **59**: 499–548.
20. Muller, F.A. and M.P. Seevinck, (2009) Discerning Elementary Particles. *Philosophy of Science* **76**: 179–200.
21. Unruh, W.G. and R.M. Wald, (1984) What Happens when an Accelerating Observer Detects a Rindler Particle? *Physical Review D* **29**: 1047–1056.
22. Weatherson, B. (2008) Intrinsic vs. Extrinsic Properties. In Edward N. Zalta (ed.), *The Stanford Encyclopaedia of Philosophy* (Fall 2008 Edition). http://plato.stanford.edu/archives/fall2008/entries/intrinsic-extrinsic.

Chapter 7
Time, Observables, and Structure

Dean P. Rickles

7.1 Mathematical Structure and Reality

Mathematics, broadly speaking, is the science of patterns. Physics, broadly speaking, is the search for patterns *in the natural world*. Eugene Wigner's [16] famous conundrum concerning the 'unreasonable effectiveness of mathematics in the natural sciences' constitutes an expression of puzzlement over the empirical success of physics, based as it is on mathematics.[1] Put this way, of course, the problem has a rather simple answer: mathematics (the science of patterns) is so effective because the natural world (the subject matter of physics) *is itself patterned*. The regularities of physics are but instances of mathematical structures. For example, we can apply geometry to physical space because physical space has a structure that is (more or less) isomorphic to some geometrical structure (or a sequence of such structures, in the case of evolving 3-geometries, or geometrodynamics). We might, in somewhat different terms (and ignoring complications to do with representation), view our world as a *model* of the axioms of some systems of geometry (and of the axioms of quantum field theory, say—though this latter connection is rather more debatable).

However, the problem is really an old one, and there are some old solutions too. Pythagoras claimed, similarly, that there was no real distinction between 'the world of physics' and 'the world of mathematics'. Plato argued there was a very great difference: it amounted to *concrete* versus *abstract*, a distinction denied by Pythagoras. For Plato, of course, the concrete, physical world instantiated (or 'partook of') the

---

This chapter was originally written for the FQXi's *Nature of Time* essay competition (http://www.fqxi.org/community/essay/winners/2008.1#Rickles). I thank FQXi for permission to reproduce the essay (albeit in modified form) here.

[1] As he puts it: 'the mathematical formulation of the physicist's often crude experience leads in an uncanny number of cases to an amazingly accurate description of a large class of phenomena' [16, p. 230].

D.P. Rickles (✉)
Unit for History & Philosophy of Science, University of Sydney, Sydney, NSW, Australia
e-mail: dean.rickles@sydney.edu.au

E.M. Landry, D.P. Rickles (eds.), *Structural Realism*, The Western Ontario Series in Philosophy of Science 77, DOI 10.1007/978-94-007-2579-9_7,

abstract forms (albeit imperfectly). This Platonic account is somewhat similar to the model-based view presented above.

More recently structural realists have answered the question about the effectiveness of mathematics by arguing that the empirical sciences are about the discovery of *structural* aspects of the world, and these structural aspects are of a mathematical nature (or, at least, of a kind that submit to mathematical representations).[2] Radical 'ontic' structural realists turn this in to an all out ontological claim: science is about structure; scientific theories encode structural information about the world; and structure is all there is. In this way, structural realists aim to provide a complete story about how and why (mathematized) scientific theories work so well.

7.2 Extreme Structuralism

Max Tegmark [14] has recently extended this basic structural realist idea by combining it with something like David Lewis' extreme brand of modal realism [6]. Not only is there the structure we observe (which is mathematical), there exist mathematical structures of all possible types! We have here, then, an extreme case of the principle of plenitude. Why would one ever wish to go to this extreme? To explain the nature and existence of the structure we observe. The laws (structure) of our universe are by no means necessary and so demand some explanation for why they are *thus* rather than *so*. Tegmark answers this question with absolute proliferation: our world is a mathematical structure in a multiverse of all possible structures. Interestingly, in one sense, though the structure we inhabit (and are, ourselves, part of) is itself contingent,[3] the existence of the structure we inhabit is in another sense *necessary*, since (being an instance of an eternal mathematical structure) it will be a possible world relative to all other worlds.

The maximal multiplicity of possible worlds, then, is utilized to ground a theory of everything that does not face problems of creation *ex nihilo*: mathematical structures are timeless, they are not the kinds of thing that can be created and destroyed. One can then, if one is so inclined, invoke an anthropic explanation of why we find ourselves in *this* particular mathematical structure. This is, admittedly, a hard view to swallow! However, if one wishes to explain *why there is something rather than nothing* (surely the ultimate explanandum?), then I see no other alternative than to propose such curious-sounding theories as the invoking of eternally existing

[2] Though not all structural realists would go this far. Some, for example, would prefer to say that structure is physical, and that there might be biological and social structures that are *not* necessarily mathematical. However, the recent trends are towards extending structural realism across the entire domain of science, including biology [5] and economics [12]. The distinction between 'physical' and 'mathematical,' if it can be established at all, is not quite as simple as it might seem prima facie—see [8] for a discussion of this difficulty in the context of string theory.

[3] For example, one can conceive of the laws being different, and indeed, as in David Lewis' theory, the existence of a plurality of structures of the sort described can provide the machinery to ground such possibilities.

structures. If we are willing to accept this, then we are led to a belief in many types of (consistent) structure. If we are then further willing to view our world as one of these structures (i.e. literally a mathematical structure), then combined with the necessary anthropics, we have an explanation of what is often considered to be an insurmountable problem.

The major problematic step (by no means the only one), I take it, is that one requiring that we believe our world (ourselves included) is a mathematical structure. All aspects of reality in our world would have to be reconceptualized in terms of such structure.[4] The common reaction to such a view bears similarities to Dr Johnson's attempted refutation of Berkeley's idealism by the kicking of a rock. In this case, the objection is that the world does not seem to be anything like a mathematical structure: mathematical structures are abstract and physical reality is concrete (whereupon you are invited to notice that there are spatiotemporally located, impermanent solid objects on which one could stub one's toe). But we have no way of knowing what it is like to be a mathematical structure: it could, after all, be just like *this*! Moreover, given our present knowledge of spacetime (on which more later), the idea that spatiotemporal location is such that it can serve to play so crucial a role as demarcating abstract from concrete seems absurd: the chronometrical structure of spacetime (including locations, conceived of as complexes of events) are themselves dynamically determined by laws of general relativity.

However, given this dynamical determination of spacetime structure, there are real problems in attempting to account for certain observed aspects of the world; time and change being cases in point. The world certainly appears to undergo change, and this, we usually (i.e. with our philosophy hats off) assume, must happen in time. How can time and change be part of a mathematical structure given that such structures are immutable and eternal? In the case of general relativity the structures appear to directly *represent* a world without time and change since the symmetries of the theory imply a zero Hamiltonian which, in the quantized theory, leads to a non-evolving Schrödinger equation.

In the remainder of this chapter I argue that recent work on problems of time and change in classical and quantum gravity can be brought to bear on the matter resulting in a satisfactory resolution. To deploy a Wheelerism, they show how one can have time without time. We can give an account (or, at least the outline of an account) of our world, *qua* mathematical structure, that at a fundamental level does not contain time. This account makes use of structuralism in a direct way. Thus,

---

[4] Related to this is the problem of equivalent mathematical structures that correspond to distinct *physical* situations. In other words, one and the same structure can be taken to represent very different systems. I don't see this to be as problematic as it is sometimes taken to be. If there are indeed differences in the physical systems, then though we *can* indeed, in many cases, represent them using the same mathematical structure (for example, the Navier-Stokes equations can be applied to all manner of prima facie very different systems), that does not thereby mean that the systems would not have some other structures more closely corresponding to them. Any physical difference would simply mean that there ought to be a structural difference too, so long as we use a fine enough resolution of the structure.

the account I present can be usefully incorporated into Tegmark's theory—or what I shall call 'ultrastructuralism'—in order to defuse a major potential problem. As a bonus, as I mentioned above, one can extract a more physically motivated account of what structure is (at least in the context of physics).

7.3 Time and Symmetry

The universe, as a single object, is usually modelled as a four dimensional structure (a Lorentzian 4-geometry).[5] This structure is naturally changeless: change happens *within* the universe, from one hypersurface (3-geometry) to another (with the time variable chosen arbitrarily on account of general covariance). At least, within our world it seems to be constructed in this way; the laws of general relativity themselves do not completely constrain the topology of space, and do not even constrain the dimensionality of the manifold.

Of course, general relativity leads us to view spacetime geometry as part of a coupled dynamical system, as something that satisfies equations of motion and co-evolves with matter and radiation. But clearly the evolution here cannot be understood in an ordinary temporal sense, unless we have at our disposal some external time parameter against which to understand it. An alternative is to attempt to concoct some 'internal' parameter from the dynamical degrees of freedom that can then parametrize the evolution. The former 'external' parameter simply doesn't make sense in the context of general relativity. The internal parameter approach does not make sense at a *global* level, but local times can be established using appropriate invariants. However, even here, the notion of systems evolving against coordinate time is inappropriate: the time that emerges is a dynamical construction from events.

This, in a nutshell, is the problem of time in general relativity: spacetime geometry is a dynamical variable, but clearly the dynamics cannot be understood in the usual sense (that is, as involving an external time parameter). The problem is worse than this, however, and can trickle down from global (involving the universe as a whole and a timelessness that is fairly innocuous) to local (involving timelessness and changelessness of the states and observables *within* the universe). At the root of this problem is the symmetry group of general relativity, the group of four-dimensional diffemorphisms of the spacetime manifold. Diffeomorphism invariance makes local observables (i.e. observables sitting at spacetime points or within regions of spactime) an impossibility, for the equations of motion (of generally relativistic theories) are invariant with respect to diffeomorphisms that shift the points and regions about. Since there clearly *are* (in some sense) localized degrees of freedom, and these are what we observe (and that seem to evolve), we need some notion of local observable that does not make reference to the spacetime

[5] To this we might add various geometric object fields representing the observed matter and radiation.

manifold but that fits our experience. That is, we need a background independent notion of observable that does not utilize external spatial and temporal parameters, for changes with respect to these will be symmetries of the theory.

A popular response is to use physical degrees of freedom to define observables and evolution. This can be understood as one kind of implementation of the 'internalist' strategy mentioned above. The observables so 'localized' are relational in the sense that they are not defined on a background space but only relative to other dynamical entities (matter fields, spatial volume, etc.). Observables are not of the form $\mathscr{A}(x, t)$ (where x and t label an independent manifold) but $\mathscr{A}(\mathscr{B})$ (where $\mathscr{B}$ is another observable and neither $\mathscr{B}$ nor $\mathscr{A}$ is privileged in any sense). One can then consider the relative evolution of such observables, looking at the way in which changes in the value of one are correlated with changes in the value of the other. This approach can give us notions of time and change that emerge as a consequence of functional relations between elements of a mathematical structure. However, this is to oversimplify matters: in order to properly appreciate the nature of this problem, and the suggested resolution[6] I need to quickly cover the entangled concepts of gauge and constraints. I can then introduce Rovelli's framework for constructing suitably invariant observables, for constrained (gauge) systems, and show how it provides a structuralist response to the problem of time that can be utilized by the ultrastructuralist to explain time in an atemporal mathematical structure.[7]

7.4 Constraints and Gauge

The problems of time and change sketched above are aspects of the fact that general relativity is a gauge theory—its Hamiltonian formulation is given by constraints. The 'gauge' here simply refers to the freedom in choosing coordinates used to parametrize space so that choosing a particular coordinate system amounts to choosing a gauge. Physical quantities are those that are independent of such a gauge choice. We give a very rough and ready presentation of these ideas here—for more details (in the context of the problem of time), see [11].

The diffeomorphism symmetry mentioned above affects the dynamics so that a standard Hamiltonian or Lagrangian formulation of the theory is not possible. Respectively, the canonical variables, q and p, are not all independent (being required to satisfy identities known as constraints: $\phi(q, p) = 0$) and the Euler-Lagrange equations are not all independent. These identities serve to 'constrain' the

---

[6] This view has been defended by a variety of authors; most notably Bryce DeWitt and Carlo Rovelli. Here I adopt Rovelli's 'partial' and 'complete' observables formalism [13]. See [11] for a general review of the problem of time and proposed solutions.

[7] I restrict the discussion to classical systems in order to make the presentation easier to follow. For the technically savvy, one can transform to the quantum case, roughly, by thinking of the functional relation or correlation $\mathscr{A}(\mathscr{B})$ as representing the expectation values of $\mathscr{A}$ relative to the eigenvalues of $\mathscr{B}$.

set of phase space points that represent genuine physical possibilities: only those points satisfying the constraints do so, and these form a subset in the full phase space known as the 'constraint surface'. This has a direct impact on the form of the observables. Since a pair of dynamical variables (not observables) that differ by a gauge transformation are indistinguishable, corresponding to one and the same physical state of affairs (the defining characteristic of a gauge transformation), the observables ought to register this fact too: that is, the observables of a gauge theory should be *insensitive* to differences amounting to a gauge transformation—as should the states in any quantization of such a theory: i.e. if x is related to y by a change of gauge, then states Ψ must satisfy the gauge condition that $\Psi(x) \approx \Psi(y)$, where we use Dirac's 'weak' equality, '$\approx$', to denote identity on the constraint surface.[8]

More explicitly, where '$\mathscr{A}$' is a dynamical variable, 'O' is the set of (genuine) observables, x, y are states (represented by points on the constraint surface), and '$\sim$' denotes gauge equivalence, we can express this as:

$$\mathscr{A} \in \mathsf{O} \iff (x \sim y) \supset (\mathscr{A}(x) \approx \mathscr{A}(y)) \tag{7.1}$$

Or, equivalently, we can say that the genuine observables are those dynamical variables that are constant on gauge orbits '$[x]$' (where $[x] = \{y : y \sim x\}$):

$$\forall [x]\ ,\ \mathscr{A} \in \mathsf{O} \iff \mathscr{A}[x] = \text{const.} \tag{7.2}$$

Most of the work done on constructing the observables of general relativity is done using the 3 + 1 projection of the spacetime Einstein equations. That is, the constraints are understood as conditions laid down on the initial data $\langle \Sigma, h, K \rangle$ when we project the spacetime solution onto a spacelike hypersurface Σ—here, h is a Riemannian metric on Σ and K is the extrinsic curvature on Σ; note that this formulation has since been superseded by a representation in terms of Wilson loops and their conjugate momenta (namely, fluxes). I won't go into the nitty gritty details here, but it turns out that the Hamiltonian of general relativity is a sum of constraints on this initial data (of the kind that generate gauge motions, namely 1st class)—hence, the dynamics is entirely generated by constraints and is therefore pure gauge. There is no evolution in time. This is the technical expression of the problems posed above and makes perfect sense if one thinks of time evolution as a diffeomorphism generating the shifting of data from one slice to another.

[8] It seems that Einstein might have been aware of this implication soon after completing his theory of general relativity, for he writes that 'the connection between *quantities in equations* and *measurable quantities* is far more indirect than in the customary theories of old' [4, p. 71] .

This formulation allows us to connect up the characterization of the observables to the dynamics (generated by constraints, abbreviated to $\mathscr{H}_i$) more explicitly:

$$\mathscr{A} \in \mathsf{O} \iff \{\mathsf{O}, \mathscr{H}_i\} \approx 0 \ \forall i \quad (7.3)$$

In other words, the observables of the theory are those functions that have weakly vanishing (i.e. on the constraint surface) Poisson brackets with all of the (first-class) constraints.[9] These are the physical, gauge-invariant quantities: evolving with the constraints (the dynamics) does not generate a physically distinct state but simply changes the gauge.

A pressing problem in general relativity—especially pressing for constructing a quantum theory of gravity—is to find suitable entities that satisfy this formal definition. There are at least two types that fit the bill, both non-local in some way: (1) global quantities defined over the whole spacetime[10] and (2) 'relational' quantities built out of correlations between field values and/or invariants. There seems to be some consensus forming that the latter type are the way to go, and these will serve as the appropriate vehicle for defining time in an unchanging mathematical structure, as well as defining the structures themselves.

7.5 Complete Observables as Structural Correlations

John Earman calls quantities of the form $\mathscr{A}(\mathscr{B})$ 'coincidence occurrences'.[11] As he explains, 'a coincidence occurrence consists in the corealization of values of pairs of (non-gauge invariant) dynamical quantities' [2, p. 16] . Earman thinks that this new conception of physical quantities signals the necessity of a shift from the traditional 'subject-predicate'-based ontologies, such as substantivalism and relationalism. I think this is the right thing to say, and have argued for and developed this point elsewhere [9–11]. It bears a striking resemblance to the version of structuralism espoused by Eddington:

> any conception of structure (as opposed to substance) must be analysable into a complex of relations and relata, the relata having no structural significance except as the meeting point of several relations, and the relations having no significance except as connecting and ordering the relata. [3, p. 121],

---

9 There are two types of constraint in general relativity: the Hamiltonian (or scalar) constraint and the momentum (or vector) constraint. These can be understood as encoding indeterminacy about 'when and where' some quantity is measured.

10 There is a proof (for the case of closed vacuum solutions of general relativity) that there can be no *local* observables at all [15], where 'local' here means that the observable is constructed as a spatial integral of local functions of the initial data and their derivatives.

11 We might also call them 'Kretschmann observables' since they stem from Kretschmann's objection to general covariance later incorporated into Einstein's own 'point-coincidence' argument.

However, I spell it out rather differently. Rovelli's framework of partial and complete observables—developed in [13]—provides, I think, the perfect formal framework in which to ground the conceptual framework.

- A *partial* observable is a physical quantity to which we can associate a measurement leading to a number
- A *complete* observable is defined as a quantity whose value (or probability distribution) can be predicted by the relevant theory.

Partial observables are taken to coordinatize an extended configuration space $\mathscr{Q}$ and complete observables coordinatize an associated (physical) reduced phase space Γ_{red}. The 'predictive content' of some dynamical theory is then given by the kernel of the map $f : \mathscr{Q} \times \Gamma_{\text{red}} \rightarrow \mathbb{R}^n$. This space gives the *kinematics* of a theory and the *dynamics* is given by the constraints, $\phi(q^a, p_a) = 0$, on the associated extended phase space $T^*\mathscr{Q}$.

The content appears to be this: there are quantities that can be measured whose values are *not* predicted by the theory. Yet the theory *is* deterministic (modulo quantum theoretic probabilities) because it does predict *correlations* between partial observables (i.e. complete observables). The dynamics is then spelt out in terms of *relations* between partial observables. Hence, the theory formulated in this way describes relative evolution of (non-gauge invariant) variables as functions of each other. No variable is privileged as *the independent one* [7, p. 5] . The dynamics concerns the relations between elements of the space of partial observables, and though the individual elements do not have a well defined evolution, relations between them (i.e. correlations) do, and in such a way as to remain independent of coordinate space and time.

The interpretation vis-á-vis time is as follows: let $\phi = T$ be a partial observable parametrizing the ticks of a clock (laid out across a gauge orbit), and let $f = a$ be another partial observable (also spanning a gauge orbit). Both are non-gauge invariant quantities. A gauge *invariant* quantity, a complete observable, can (here borrowing from [1]) be constructed from these partial observables as:

$$\mathscr{A}_{[f;T]}(\tau, x) = f(x') \tag{7.4}$$

These quantities encode correlations. They tell us what the value of a non-gauge invariant function f is when, under the flow with respect to the vector field generated by the constraint, the non-gauge invariant function T takes on the value τ. This correlation is gauge invariant. These are the kinds of quantity that a background independent gauge theory like general relativity is all about. We don't talk about the value of the gravitational field *at a point of the manifold*, but where some other physical quantity (say, a value of the electromagnetic field) takes on a certain value. In this sense, this formal framework codifies Eddington's earlier views, and also Earman's characterisation of coincidence occurrences.[12] I prefer to think of the complete observables as primitive structural correlations that are reducible only in unphysical ways (corresponding to distinct gauge choices).

7.6 Relations Without Relata

Let us return to the issue of structuralism. Epistemic structural realists argue that the best we can hope for is to get to know structural aspects of the world, since we only ever get to observe relational rather than intrinsic properties (in our experiments and so on). However, in a background independent gauge theory like general relativity we have seen that the physical observables just *are* relational quantities: this is all there is! One cannot even speak of an independent spacetime. Hence, the notion of a non-relational quantity, defined at a point of spacetime, is physically incoherent, though we can provide a purely *formal* expression of the notion in terms of gauge variant coordinate-dependent quantities. As an example we can think of a GPS [Global Positioning System] observable, involving the measurement of some field component at a location determined by a physical GPS coordinate system.

In other words, there's nothing 'underneath' the relational properties (as encoded in the 'overlapping' dynamical fields), so that these *exhaust* what there is, leading to an *ontological* structuralism motivated entirely by the proper conceptualisation (and formalisation) of general relativity. Hence, we have here an empirical argument for ontic structural realism that evades the standard 'no relations without relata' objection. The relations are the correlations here (the gauge invariant, complete observables), and the 'relata' would be the non-gauge invariant, partial observables. But the partial observables being non-gauge invariant do not correspond to physical reality: only the complete observables do. Partial observables correspond to an arbitrary choice of gauge that can be transformed away. We cannot *decompose* the correlations in an ontological sense, though we clearly can in a epistemic or formal sense—indeed, the correlates constitute our 'access points' to the more fundamental, physical correlations.

We talk about correlations in terms of quantities that are correlated. But there is a clear ontological division between the status of partial observables (i.e. relata) and complete observables (i.e. relations).[13] This is, then, precisely why we face problems regarding the 'subject-predicate'-style ontologies that Earman mentions: there

[12] Once again, we find that Einstein was surprisingly modern-sounding on this point, writing that 'the gravitational field at a *certain location* represents nothing "physically real", but the gravitational field together with other data does' [4, p. 71] . Likewise, the 'other data' will represent nothing without yet more data (such as the gravitational field). The correlations are the fundamental things.

[13] Here then we have a clear answer to Chakravartty's question 'in what sense are relations more fundamental?' (Chapter 10, p. 199). The physical weight is carried by the complete observables (the relational structures), and the partial observables (relata) are certainly not fundamental in the sense that they are not even physical but artefacts of the gauge choice employed.

are no independent subjects that are the 'bearers' of properties and the 'enterers' of relations. We can seemingly invoke such subjects in our representations, but it is mere artifice.

What is interesting about this version of ontic structural realism is its independence from standard philosophy of science issues. We do not need to ground the position in the underdetermination of theory by data or the pessimistic meta-induction. The method of motivation is entirely internal to the interpretation of mature physical theory providing an alternative path to structural realism. Of course, there are many issues still missing from this brief account, not least how to think about causality, laws, and such like. However, it does indicate one possible path to ontic structural realism based in *real* physics as opposed to promissory notes.

7.7 Conclusion

The position I have described involves the idea that physical systems (which I take to be characterized by the values for their observables) are exhausted by extrinsic or relational properties: they have no intrinsic, local properties at all! This is a curious consequence of background independence coupled with gauge invariance and leads to a rather odd picture in which objects and structure are deeply entangled. Inasmuch as there are objects at all, any properties they possess are structurally conferred: they have no reality outside some correlation. What this means is that the objects don't *ground* the structure; they are nothing independently of the structure, which takes the form of a (gauge-invariant) correlation between (non-gauge invariant) field values. With this view one can both evade the standard 'no relations without relata' objection and the problem of accounting for the appearance of time (in a timeless structure) in the same way.

References

1. Dittrich, B. (2007) Partial and Complete Observables for Hamiltonian Constrained Systems. *General Relativity and Gravitation* **39**: 1891–1927.
2. Earman, J. (2006) The Implications of General Covariance for the Ontology and Ideology of Spacetime. In D. Dieks (ed.), *The Ontology of Spacetime* (pp. 3–23). North Holland: Elsevier.
3. ——— (1921) A Generalisation of Weyl's Theory of the Electromagnetic and Gravitational Fields. *Proceedings of the Royal Society of London. Series A* **99**(697): 104–122.
4. Einstein, E. (1918) Dialogue about Objections to the Theory of Relativity. In A. Engel (ed.), *The Collected Papers of Albert Einstein* (pp. 114–122), *Vol. 7, The Berlin Years: Writings: 1918–1921, English Translation of Selected Texts*. Princeton, NJ: Princeton University Press, 2002.
5. French, S. (2011) Shifting to Structures in Physics and Biology: A Prophylactic for Promiscuous Realism. *Studies in the History and Philosophy of Biology and the Biomedical Sciences* **42**(2): 164–173.
6. Lewis, D.K. (1986) *On the Plurality of Worlds*. Oxford: Blackwell.

7. Montesinos, M., C. Rovelli, and T. Thiemann (1999) An $\mathsf{SL}(2, \mathbb{R})$ Model of Constrained Systems with two Hamiltonian Constraints. *Physical Review D* **60**: 044009.
8. Rickles, D. (2010) Mirror Symmetry and Other Miracles in Superstring Theory. *Foundations of Physics*. Online First: DOI: 10.1007/s10701-010-9504-5.
9. Rickles, D. (2008) Who's Afraid of Background Independence. In D. Dieks (ed.), *The Ontology of Spacetime II* (pp. 133–152). North Holland: Elsevier.
10. Rickles, D. (2007) *Symmetry, Structure, and Spacetime*. North Holland: Elsevier.
11. Rickles, D. (2006) Time and Structure in Canonical Gravity. In D. Rickles, S. French, and J. Saatsi (eds.), *The Structural Foundations of Quantum Gravity* (pp. 152–195). Oxford: Oxford University Press.
12. Ross, D. (2008) Ontic Structural Realism and Economics. *Philosophy of Science* **75**: 731–741.
13. Rovelli, C. (2002) Partial Observables. *Physical Review D* **65**: 124013–124013–8.
14. Tegmark, M. (2006) The Mathematical Universe. *Foundations of Physics* **38**: 101–150.
15. Torre, C. G. (1993) Gravitational Observables and Local Symmetries. *Physical Review D* **48**: R2373–R2376.
16. Wigner, E. P. (1967) The Unreasonable Effectiveness of Mathematics in the Natural Sciences. In E. P. Wigner (ed.), *Symmetries and Reflections: Scientific Essays* (pp. 222–237). Bloomington: Indiana University Press.

Part III
Modality and Causality in Structural Realism

Chapter 8
Ontic Structural Realism and Modality

Nora Berenstain and James Ladyman

8.1 Introduction

Ontic structural realism (OSR), as originally developed by Steven French and James Ladyman, incorporates an explicit commitment to the claims that the world has an objective modal structure and that this structure is represented by the theoretical structure of our best scientific theories. OSR attributes a rich structure of natural necessity to the world, and this distinguishes it from Bas van Fraassen's structural empiricism as well as from John Worrall's epistemic structural realism, both of which reject natural necessity. Ladyman and Ross [12, 13, 21] extend OSR from physics to the special sciences and invoke the notion of the objective modal structure of reality in their accounts of novel prediction, causation, and real-patterns ontology. One may think about the modal structure discovered by science primarily in terms of causes or in terms of laws of nature. Either way, it is reasonable to question the extent to which the commitment to objective modality is essential to the defense of OSR and whether enough has been done by the above authors to justify its inclusion in their avowedly naturalistic metaphysics.

It is instructive in this context to consider whether ordinary scientific realism requires a commitment to natural necessity. While there has been surprisingly little direct discussion of this question in the realist literature, it is an implicit issue throughout the debate about scientific realism and an explicit one in debates about causation, explanation, and laws of nature. David Armstrong, Alexander Bird, Nancy Cartwright, Anjan Chakravartty, Brian Ellis, Ron Giere, and Ernan McMullin are all realists and anti-Humeans. However, Barry Loewer, David Papineau, and Stathis Psillos argue for the combination of scientific realism and Humeanism. On the antirealist side, it is natural to associate the denial of scientific realism with empiricism and the rejection of necessary connections in nature. This view motivated the logical positivists and logical empiricists in their critiques of realism, and it was also the view held by their intellectual ancestors. More recently, van Fraassen

---

N. Berenstain (✉)
University of Texas, Austin, TX, USA
e-mail: nberenstain@mail.utexas.edu

E.M. Landry, D.P. Rickles (eds.), *Structural Realism*, The Western Ontario Series in Philosophy of Science 77, DOI 10.1007/978-94-007-2579-9_8,

has repeatedly argued that scientific realism requires modal metaphysics and is for this very reason to be rejected by empiricists. (Of course, the above-mentioned Humean scientific realists will deny both of his claims.) On the other hand, it is clearly possible to deny important components of scientific realism and still be a realist about modality in other contexts. Non-naturalists, for example, might privilege agency and the manifest image in such a way as to be anti-Humean in their metaphysics while also making the repudiation of the scientific image central to their philosophy.

In this paper, we argue that not only is a reliance on objective modality integral to arguments for scientific realism but so too is a commitment to natural necessity. We also argue that the most popular and viable form of Humean scientific realism faces serious difficulties and therefore that the scientific realist must embrace anti-Humeanism about laws and causation. Before delving into these arguments, however, the notion of objective modality must be clarified. In accordance with methodological naturalism, we formulate our notion of objective modality not simply in the abstract but with reference to actual science. Consider, for instance, the cosmic speed limit of the speed of light. That material bodies do not in fact exceed c is not what is usually claimed; rather it is said that material bodies *cannot* go faster than light. It is also generally pointed out that it would take an infinite force to accelerate an ordinary massive object to the speed of light, because in any frame in which the body were traveling anywhere close to light speed it would be unimaginably heavy. The question is whether modal claims such as these have their truth-values independently of our current theories and models. If they do, then the modal facts are objective in the sense that they do not depend on our doxastic or epistemic states. If we think in terms of laws, then the question is whether the laws of nature are what they are because of how the world is independently of the way we as conscious subjects describe it. The matter may also be put in terms of causation. In the idiom of contemporary metaphysics, we may ask whether there are genuine causal powers.

Of course, it is possible to deny the existence of natural necessity while allowing that the modal language of science captures the way the world is independently of us. For example, it is possible to give a reductive account of modal facts that does not reduce them to anything that is dependent on us. This means that it is not correct to contrast Humeanism with the acceptance of objective modality. Rather, Humeanism is opposed to *irreducible* objective modality, or any kind of primitive natural, nomological, or causal necessity.[1] Humeans deny that there is natural necessity but may nonetheless offer an account of objective modality. The sophisticated Humean may contend that the realist arguments for natural necessity depend only on there being objective modality, for which natural necessity is sufficient but not necessary. Hence, he will argue, Humeans can have the best of both worlds and avail themselves of the explanatory and dialectic advantages of scientific realism without incurring the metaphysical cost. We deny the tenability of this position.

---

[1] We take natural necessity to be that to which one is committed when one accepts realism about laws of nature as something over and above regularities among actual events.

In the next section, we review scientific realism and argue that realists are prima facie committed to objective modality and that the main argument for realism deepens that commitment. In Section 8.3, the most defensible form of Humeanism in the contemporary literature is introduced and criticized as being incompatible with the spirit of scientific realism. Section 8.4 is a brief conclusion.

8.2 Scientific Realism and Objective Modality

Historically, there has been a connection between those who have criticized scientific realism and those who have opposed metaphysics of all kinds, particularly modal metaphysics. Scientific realists, on the other hand, are often distinguished from one another by their particular ontologies of causes, laws, and properties. However, if we define scientific realism simply as the view that our best scientific theories commit us to the existence of certain unobservable entities, then the position need not entail the existence of natural necessity. Rather, the modal implications of scientific realism come from the arguments in favor of the position and from the way its advocates articulate their view of science. In any case, it is not yet clear what it means to say that scientific realism is committed to the existence of natural necessity or simply to objective modality of some sort, as there is no widespread agreement on what counts as scientific realism. Let us review some of the relevant controversial propositions that are taken to characterize the view:

1. Many or at least some of our best scientific theories and laws are true (or approximately true) claims about the world.
2. Many or at least some of the theoretically described unobservable entities that we talk about exist.
3. Many or at least some of the theoretical terms in our best scientific theories genuinely refer to unobservable entities.
4. The aim of science is truth, not mere empirical adequacy.
5. The best explanation of the success of science is that our theories are correctly describing the unobservable causes of what we observe.

1 implies a commitment to laws but not necessarily to causation. It is important to note that philosophers of science are divided over whether causation is a fundamental feature of reality, following Russell's famous claim that it does not feature in advanced sciences such as fundamental physics. In any case, 1 does not make it explicit whether laws are to be understood in genuinely modal terms or just as generalizations over regularities. The Stanford school of philosophers of science, notably Nancy Cartwright and Ian Hacking, developed entity realism and assert 2 but deny 1. Obviously, they are not ontologically committed to objective modality in virtue of some commitment to laws of nature. However, they do invoke a non-Humean notion of causation, and indeed Cartwright [3] is a sustained attack on Humeanism about causation, which proposes to replace the view with a metaphysics

of causal capacities. Here, the commitment to objective modality stems from the implicitly modal nature of the notion of capacity.

The reference of theoretical terms has been much discussed both in the abstract as well as in concrete scientific cases. The causal theory of reference was central to Hilary Putnam's post-positivist development of scientific realism and is still central to many scientific realist responses to the pessimistic meta-induction, such as [9]. This is an example of how objective modality may be a commitment incurred by some realists not in virtue of their realism but in virtue of their arguments for it (see Section 8.2.2 below).[2] While the distinction between truth and empirical adequacy may seem modally neutral, note that it is often spelled out in terms of the distinction between explaining the phenomena and merely describing them. Note also that it may be argued that even the notion of empirical adequacy is modally committing insofar as it pertains to a description of the observable phenomena and not merely the actually observed phenomena.[3] 5 is one form of the argument for scientific realism based on inference to the best explanation, where this is construed as a causal explanation of the observable in terms of the unobservable.

8.2.1 Prima Facie *Considerations: Modal Notions in Science*

In rough terms, realists are those who take the theoretical claims of science at face value and who regard scientific theory choice as the final word in the epistemology of particular scientific theories and statements. Realism requires rejecting both revisionary philosophical versions of scientific theories and philosophical arguments for any skepticism that goes beyond the scientific community's healthy epistemic caution, exemplified by its insistence on empirical testing. If theoretical claims about electrons are to be taken literally as referring to unobservable entities bearing certain properties, then so too should claims about laws, causes, and other modalities. The following brief tour through modal notions in science is intended to establish that the skeptic about objective modality in general and natural necessity in particular faces the burden of proof and must offer some positive account that makes sense of the ubiquity of modal notions in science.

8.2.1.1 Laws

The pursuit of laws of nature is so central to the practice of science that establishing a body of laws concerning some domain is often considered to be a necessary condition for the formation of a science. Consequently, it has been argued that economics and other social sciences are not properly scientific because of the paucity

[2] Psillos [19] develops a hybrid causal descriptivist theory of reference and applies it to various cases of abandoned theoretical terms. While he is officially a Humean, there is a question as to whether the causal theory of reference is viable if one denies that causation may be a singular relation between language users and events.

[3] See [11] and the reply by Monton and van Fraassen.

of empirically confirmed laws in their domains. Both critics and advocates of social science have claimed that it is at least fundamentally different from natural science because of its deficit of laws. The status of physics, on the other hand, is due in large part to its success in reducing the complexity of nature to the operation of a few fundamental laws, beginning with Newton's three laws of mechanics and law of universal gravitation and culminating in the symmetry principles and conservation laws catalogued in the last century. Popular and semi-popular debates among physicists about the origins of the universe and the so-called 'fine-tuning problem' make reference to strong distinctions between contingent facts like the values of free parameters in the Standard Model and necessary facts about what the laws of nature allow. The necessitarian view of laws of nature is articulated in bold terms by David Armstrong and Fred Dretske, who argue that laws of nature are necessary relations between universals. Other anti-reductionist views of laws include those advocated by John Carroll and Marc Lange.

Laws, however, are not celebrated only by realists and metaphysicians. Carnap criticized organicist biology as pseudo-science, because it lacked laws and so the precision and quantitative prediction that he thought were essential to true science. However, he rejected the idea that laws had any modal status and thought of them as simply maximally general statements. Empiricists in general may prize laws for their systematizing of the phenomena but deny that they do more than describe how things behave. Kepler's laws of planetary motion, like many phenomenological laws, seem readily understood as descriptive rather than explanatory, and are not usually stated in modal form. But more fundamental laws are very often formulated modally. For example, the first law of thermodynamics (the law of conservation of energy) states that energy *cannot* be created or destroyed. Pauli's exclusion principle says that two electrons *cannot* have the same set of quantum numbers and that fermionic wavefunctions *must* be antisymmetrized. It is even possible to find examples of less fundamental laws that seem modally committing. In chemistry, enantiomers, i.e. stereoisomers related by a reflection, *cannot* be superimposed on one another. In Section 8.3, we consider in depth the matter of the modal status of laws.

8.2.1.2 Causation

Scientific journals and textbooks are awash with causal claims, and much important scientific knowledge takes the form of the attribution of causal powers to objects. However, this did not stop Bertrand Russell from famously denying that science dealt with causes, and there has been a recent revival of his causal republicanism [18]. Similarly, causation was subordinate to laws in the work of logical positivists and logical empiricists, whose view of science focused on generality and universality, rather than the particular and the local. As mentioned above, the Stanford School and especially Nancy Cartwright combined their skepticism about high theory and universal laws with a robust realism about singular causation as attributed to events and processes in the laboratory. The Humean theory of causation focuses on generic causal relations between types of entities and events. In doing so, Cartwright argues, it abstracts away the complexity and specificity of concrete phenomena and misrepresents their nature. The metaphysics of causal powers that she developed is designed to draw our attention back to the way that science works in practice by attributing more or less stable causal powers to things, powers that can reinforce or inhibit each other, and whose combined action is a local outcome of singular interactions rather than the diktat of inviolable law.

Even those who do not follow Cartwright in prioritizing causal powers over lawhood are often at pains to explain how causal modality is objective albeit derivative of lawlike necessity. Armstrong, for example, argues that there is a singular causal relation between instances of properties just in case the properties in question are instantiations of universals between which there are necessary connections.

8.2.1.3 Probability

Probability was on Aristotle's list of modalities with good reason. Like other modal operators, such as the actuality, necessity, possibility, and potentiality operators, the operator 'probably' has the property of being non-truth-functional. Probability also seems to create intensional contexts. For instance, it was true in 2005 that it was improbable that the next president of the United States would be an African-American, but it was not true that it was improbable that Obama would remain an African-American. The connections between probability and potentiality are complex. On one hand, probability seems to be a kind of quantification of the potential for something to be the case. For example, there is the potential for one to be dealt four aces in a hand of five cards but not as much potential as there is to be dealt three twos. On the other hand, we use potentiality to relate finite frequencies to probabilities. Hence, after tossing a fair coin ten times and obtaining seven heads and three tails, we suppose that if we were to continue tossing the coin the relative frequencies would approach one half. While there are no actual infinite trials, we suppose that in a series of finite trials of increasing lengths the relative frequency would tend to the probability in the limit.

It is common to distinguish epistemic probabilities from objective chances. As the former may be treated in common with other epistemic modalities, it does not pose a challenge to the antirealist about objective metaphysical modality. Thus it is objective chances, or probabilities construed as real features of the world and not merely as features of our beliefs, that concern us here. David Lewis constructs a cosmology that purports to maintain objective chance despite placing very few constraints on which possible worlds are accessible from any given world.[4] Other options include taking single-case chances to be a brute feature of the world, as in the theory of propensities, or somehow reducing objective chance to a result of the modal force of indeterministic laws. Prima facie the difference between a deterministic and an indeterministic theory must itself be a modal one,

[4] Barry Loewer's theory of probability is a development of Lewis's.

since any given sequence of actual events could be produced by either a deterministic or a stochastic process.

It may be that probability can be understood without metaphysical commitments to anything other than concrete events in the actual world, but it is certainly not obvious that it can be. Without question, probability is now absolutely essential to scientific theorizing and experimentation. It is of great foundational significance in statistical mechanics and quantum field theory, as well as in decoherent-histories approaches to cosmology. It is with good reason that van Fraassen calls probability the new modality of science and admits that, as an empiricist who embraces probabilism in epistemology, his own modal-frequency account of probability is what brings him closest to metaphysics.

8.2.1.4 Equilibrium

Equilibrium explanation is extremely important in evolutionary biology and holds great promise in the social sciences because of the development of evolutionary game theory. It is a different brand of scientific explanation than causal explanation:

> Where causal explanation shows how the event to be explained was in fact produced, equilibrium explanation shows how the event would have occurred regardless of which of a variety of causal scenarios actually transpired. [23, p. 201]

As with causal explanation, however, science's use of equilibrium explanation carries a prima facie commitment to the notions of physical necessity and possibility.

As an example of equilibrium explanation, Elliott Sober considers R.A. Fisher's 1931 account of why the sex ratio is 1:1 for many species. Fisher offers the explanation that if a population ever departs from this ratio, there will be a reproductive advantage favoring parental pairs that overproduce the minority sex. Supposing that males are in the majority of a given species, parental pairs that produce all female offspring will have greater fitness than pairs that produce a mix of male and female offspring, where fitness is measured in terms of number of grandchildren produced. This advantage will be retained until the species once again reaches the equilibrium sex ratio of 1:1.

It is clear that equilibrium explanation is in some sense broader than causal explanation. A causal explanation of the sex ratio for a given population would cite earlier conditions of the population, as well as evolutionary forces that move the population to the equilibrium state. The equilibrium explanation, however, need not refer to the population's earlier state, as it explains the population's equilibrium state *regardless* of its earlier states. Sober writes, 'Whatever the actual initial sex ratio *had been*, the selection pressures that would have resulted *would have* moved the population to its equilibrium state' [23, p. 202]. The equilibrium explanation works for any population in equilibrium because it explains why, given any previous non-equilibrium state, the population *must* tend toward equilibrium. It seems then that equilibrium explanation appeals to natural necessity to account for systems being in equilibrium states.

8.2.2 The No-Miracles Argument

The main motivation for scientific realism in general and structural realism in particular is the no-miracles argument. The no-miracles argument is a form of inference to the best explanation. Many of those who advocate the use of IBE argue that it is only when scientific hypotheses latch onto modal structure that they are genuinely explanatory. For example, it is because the atomic structures of elements determine their possible patterns of interaction that the full valence-shell configuration of the noble gases explains why such gases tend not to participate in chemical reactions. The scientific realist argues that novel predictive success gives us a reason to believe in the unobservable entities posited by our best scientific theories. The connection between novel predictive success and unobservable entities can be understood as follows. Unobservable entities have properties and obey laws that successful theories more or less correctly describe. Hence, such theories can predict in advance what the behavior of these entities will be in novel situations and what phenomena the behavior in question will cause.

Scientific explanation creates intensional contexts without explicitly invoking either laws or causes. Like law statements and causal claims, explanations are generally not preserved under substitution of co-referential terms. For example, the temperature of the oceans is rising because of the increase in the amount of carbon dioxide in the atmosphere but not because of the increase in the atmosphere of the amount of the gas first isolated by Priestley. (Of course, it is only in the light of background theory that the former explanation explains anything at all). So the realist's reliance on explanation brings with it prima facie modal commitments.

It is also the realist's reliance on explanation that makes it hard to see how a Humean scientific realist can consistently motivate both of these positions. Antirealists reject the desire for explanation that motivates realism. Van Fraassen , for instance, regards all regularities in the phenomena as brute and denies that novel predictive success requires any explanation over and above the empirical adequacy of the theory in question. Since the scientific realist distinguishes herself from the antirealist in part by her commitment to explanation, the realist who repudiates causal necessity is in an awkward dialectical situation. If novelly predicted phenomena are not genuinely caused by the relevant unobservable entities, then in what sense do the latter explain the former? There are many things that are correlated with one another, but the asymmetric relation of explanation does not hold between them unless they are also causally or lawlike related. It is unclear how a realist who disavows natural necessity can make sense of the idea that unobservable entities explain the phenomena. The argument that we should believe in atoms because, if they exist, they are *correlated* with the phenomena is no more satisfying than the antirealist's denial that regularities ever require explanation.

8.3 Against Humeanism

Humeanism about causation is the denial that there is any such thing as causal, nomological, or natural necessity. Put another way, it is the view that there are no necessary connections in nature. For the Humean, causal connections and laws of nature are just a species of regularity among events. Of course, in everyday life and in scientific theory and practice, not all regularities are taken to be causal or lawlike. Some regularities are merely accidental patterns of correlation. Contrast, for instance, the true statement that 'all the coins in the cash register are quarters,' with the true statement that 'every planet's orbit is an ellipse.' Both statements describe regularities in nature, but only the latter is a law of nature. So statements of laws of nature are not just statements of regularities; they have some additional property, namely the property of *lawlikeness*. It is this property that enables statements of laws to support counterfactual conditionals and to figure in scientific explanations. The challenge for the Humean is to identify the property of lawlikeness, and so distinguish laws of nature from accidental regularities, without appealing to a notion of physical or causal necessity. If successful, the Humean can then claim that there is objective modality without incurring dubious metaphysical commitments.

8.3.1 The Naïve Regularity View and Epistemic Accounts of Laws

The first Humean account of laws was the naïve regularity view, which simply equated laws with universal regularities. On this view, a lawlike statement is any true universally quantified statement of the form 'all Fs are Gs.' As it denies that universal regularities must have some additional property in order to be laws, the view cannot distinguish laws from accidental regularities. This view was part of the early logical empiricist tradition and is now nearly universally rejected.

An early version of Humeanism about laws that does countenance a distinction between laws and accidents is the epistemic view, which characterizes the property of lawlikeness in terms of our particular epistemic attitude toward those statements that we take to be laws. Some version of the view has been defended by Strawson [24], Ayer [1], and Goodman [8]. On this view, laws are those generalizations that play a certain epistemic role: 'they are believed to be true, and they are so believed because they are confirmed by their instances, and they are used in proper inductive reasoning' [19, p. 141]. One problem for this view is that some laws lack any positive instances. Newton's first law, for example, concerns the motion of bodies on which no forces are exerted. Since there are no such bodies, Newton's first law cannot be a law in virtue of its confirming instances.

Another disadvantage of the epistemic view is that it does not explain *why* certain generalizations should feature in proper inductive reasoning. Since the view defines lawlike regularities in part by the role they play in induction, little in the way of explanation can be offered for why some regularities rather than others play such a role. Because the lawlikeness of a regularity depends on our epistemic attitude toward it, the view has also been criticized for the seemingly subjective and contingent nature of this criterion. It seems that we could have had different epistemic attitudes than we actually do toward any number of generalizations. Indeed, our epistemic attitude toward a specific generalization may change over time. The epistemic view has the consequences that the laws of nature change with our epistemic attitudes and that they cannot be discovered any more than customs can be. Since the idea that laws are objective, unchanging, and importantly distinct from accidental regularities is what motivated the abandonment of the naïve regularity view, the epistemic view will not detain us further.

8.3.2 The Objective Humean Account

The Mill-Ramsey-Lewis view, or as Psillos calls it, the *web-of-laws* view, is by far the most widely accepted version of Humeanism. It offers an objective characterization of lawlikeness, and so avoids the subjectivity of the epistemic view. In addition to its namesakes, the view counts Earman [6], Loewer [16], Psillos [19], and Cohen and Callender [4] among its defendants. While all but Earman identify themselves as scientific realists, we show the view to be plagued by several problems, each of which makes it difficult to reconcile with the spirit of realism.

On the MRL view, laws of nature are those regularities that feature as axioms or theorems in the best axiomatic deductive system that describes our universe. If there is one such system that can be said to offer a true description of our world, then there are many such systems. To avoid the problem of arbitrariness, Ramsey [20] and Lewis [14] both suggest that the best deductive system should be the one that strikes the best balance between simplicity and strength. For Lewis, if two or more systems tie in terms of simplicity and strength, the laws of nature are to be identified with the regularities that are common to all systems. Psillos formalizes the thesis as follows:

> (W): It is a law that all Fs are Gs iff (i) all Fs are Gs, and (ii) that all Fs are Gs is an axiom or theorem in the best deductive system ϕ (or if there is no unique best deductive system ϕ, it is an axiom or theorem in all deductive systems that tie in terms of simplicity and strength).

If there are no regularities common to all such systems, then the laws of nature are indeterminate. Apart from this problem of potential indeterminacy, we argue that the view is flawed in three significant ways: its notion of strength cannot be formulated non-circularly,[5] it can offer no explanation of the success of inductive inference, and it is not clear that it is compatible with a satisfactory account of the laws of the special sciences.

[5] While a number of authors have concerned themselves with problems for the MRL account stemming from the notions of simplicity and strength, we are aware of none that have specifically addressed this problem of circularity. See [4].

8.3.2.1 The Circularity Objection

On the web-of-laws view, we look to the best deductive axiomatized descriptions of our world to tell us which regularities count as laws of nature, and it is the theoretical virtues of simplicity and strength that tell us what those systems are. Formulating a definition of a system's simplicity is not without its difficulties, but it is the problematic notion of strength that we focus on here. Intuitively, strength has something to do with how much information can be derived from the axioms. One could say that every increase in the amount of information that is derivable from a system counts as an increase in that system's strength. This will not do, however, as any attempt to precisify such a notion seems to lead to incommensurability among systems.

Let us consider ways to explicate the intuitive notion that a system's strength is related to the amount of information derivable from its axioms. A first pass would be to formulate strength in terms of the number of facts that are derivable from a system's axioms, but this is clearly unsatisfactory. Since every system will make infinitely many facts derivable, all systems will turn out to be equally strong. Another possibility is to invoke the subset relation in the definition of strength. We could say that system A is stronger than system B iff the set of facts derivable from B form a proper subset of the set of facts derivable from A. But this notion of strength makes systems incommensurable with one another, as there will be many pairs of systems such that no subset relation holds between their respective sets of derivable facts. Is there a way to cash out the notion of strength that does not make systems either trivially equal or incommensurable in strength?

Psillos's alternative approach (though he does not state this explicitly) ties a system's strength to how well it vindicates our intuitions about what statements count as laws. While this strategy escapes the problems associated with formalizing the notion of strength, it is not without its own methodological difficulties. Psillos wants to exclude increases in a system's extraneous information (i.e. information about accidents and coincidences) from counting as increases in the system's strength. But if this is to be a viable method, there must be some principled way to distinguish between the derivable information that bears on a system's strength and the information that does not. Call this the *principle of informational relevance*. The web-of-laws proponent must be able to identify such a principle if he wishes to employ strength as a determining feature of the best axiomatic system. Since strength determines (in part) which deductive system is the best one, and that in turn determines which regularities are laws of nature, the definition of strength cannot invoke the distinction between accidental and lawlike regularities, as that is the very distinction it is in the service of explaining.

We argue that no principle of informational relevance, and therefore no tenable definition of strength, can be formulated without presupposing an established fact of the matter about which regularities are laws of nature. Consider Reichenbach's statement that there are no gold chunks that are larger than a cubic mile. Contrast it with the statement that there are no chunks of plutonium that are larger than a cubic mile. Psillos takes it as a datum that the latter but not the former expresses a law of nature.[6] Thus, the latter but not the former should be derivable from the best deductive system. If we are choosing between two (sets of) deductive systems that are the same except that one includes the former regularity as an additional axiom, clearly we should choose the system without the additional axiom. How can the web-of-laws view justify this choice? The latter system ranks higher in simplicity; but can we not also say that the former rates higher in strength? After all, the former system enables us to derive a regularity that we are not able to derive from the latter—that all gold cubes are smaller than a cubic mile. One wants to justify the choice by saying that the more complicated system is *not* stronger, but in order to do this a principle of informational relevance is required.

To further illustrate the problem, consider the following passage in which Psillos asserts that the system with the additional axiom is *not* stronger but offers nothing in the way of explanation:

> One could, of course, just add all the accidental generalizations as extra axioms to the best deductive system of the world. But in doing this, one would make this system far more complicated than it should be. If, for instance, we were to add to the best system Reichenbach's regularity that all gold cubes are smaller than one cubic mile, we would detract from its simplicity without gaining in strength. [19, p. 151]

Why wouldn't the resulting system be one that has gained in strength despite losing in simplicity? The only plausible reason is that the additional regularity we derive from the amended system is not a *law*. This response exposes the circularity inherent in Psillos's approach to explicating the notion of strength in the web-of-laws account. On this view, laws are supposed to be just those regularities that are derivable from all true deductive systems that strike a balance between simplicity and strength. But, either systems are trivially equal or incommensurable in their strength, or there must already be a fact of the matter about exactly which regularities are laws and which are accidents in order for there to be a fact of the matter about which system is stronger. Since no unproblematic notion of strength can be defined, the view's criterion of lawlikeness does not get off the ground.

8.3.2.2 Laws in the Special Sciences

Apart from its problem of circularity, the web-of-laws approach faces a lack of genuine naturalistic motivation. It is not clear that we have any good reason to believe that all or even most of what we now take to be laws of nature would be the sorts of things that would find a place in a deductive axiomatic description of the world. Psillos seems to think the web-of-laws view is right in line with the practice of science, though he gives no explanation for this. He claims, 'The useful fiction of an ideal deductive system of the world is not very far from the practice of science as we know it, nor far from what we now take the laws of nature to be.' We are not convinced.

---

[6] It is worth noting that, in his explanation of why this is the case, Psillos appeals to the notion of physical impossibility, which is circular for the Humean, as he explicates the notion of physical impossibility in terms of what is incompatible with the laws.

It may be true that attempts at unification within fundamental physics can be described as attempts to unify physics within a deductive axiomatic framework. Contemporary particle physics has been unified (for the most part) within the mathematical framework of the Standard Model, which is fundamentally a collection of symmetry groups. Mathematical systems count as deductive axiomatic systems, and so the web-of-laws view is prima facie compatible with the way fundamental physics is done. However, it is not as clear that this view can extend to the laws and practices of the special sciences.

It seems that the web-of-laws view requires all non-fundamental laws of nature to be derivable from the fundamental ones and implies that all the fundamental laws will be laws of physics. One of Psillos's footnotes supports this interpretation. He finds it worth noting that, 'Carnap also took the laws of nature to be whatever lawlike statements are *deducible* from a set of axioms that express a certain physical theory, or more generally, "the deductive system of physics" [19, p. 303]. This suggests that the web-of-laws adherent expects all laws of nature that are not themselves fundamental to be derivable from an axiomatized physics.

The difficulty with this approach is that the program of reducing special sciences to physics has often been attempted though never clearly achieved. We have little positive reason to believe that the laws of special sciences are derivable from fundamental physics, and we do have reason to believe they such derivations are computationally intractable. Our view is that the laws of the special sciences allow us to make relatively accurate predictions about macroscopic systems in the world without calculating what is happening within such systems at the minute scale of particle physics. Deriving predictions about macroscopic systems from calculations made at the scale of particle physics would require far more energy than there is in the known universe. By sacrificing some level of exactness, special-science laws allow us to make predictions that would be otherwise intractable if computed using only the fundamental laws of physics.

Consider an analogy to the famous cellular automaton, Conway's Game of Life. The 'game' is played on an infinite grid in which each cell can have one of two values, black or white. As in any cellular automaton, time is discrete and there are a finite number of deterministic update rules. Given any initial conditions—any starting distribution of black and white cells—the four update rules determine exactly what the grid will look like at every later point. Something that has fascinated Game of Life enthusiasts is the 'emergence' of macroscopic objects and patterns that seem to follow laws of their own. These higher-level laws quantify over certain types of cell configurations that behave in orderly, uniform ways. What is particularly interesting is that the derivation of many of these macro laws of the Life universe from the four underlying update rules is not a computationally tractable problem. These higher-level laws allow the Life user to make calculations about the future states of the macroscopic patterns by vastly compressing the relevant data.[7] It is our view that special-science laws, like the higher-level laws of the Life universe, can

[7] For a greater discussion of the Life universe, see [5].

be understood as tools of data-compression. A more detailed account of the view can be found in ‘Rainforest Realism and the Unity of Science’ in [13].

The important point is that, on this view, it is characteristic of special-science laws that their derivation from the underlying laws of physics requires carrying out unmanageably large computations. Even if some of what we now take to be the laws of the special sciences are, in the strict logical sense, derivable as theorems from the axioms of fundamental physics, it is not the case that we can ever know which ones these are. Thus, if the MRL view requires all non-fundamental laws to be deducible from the axioms of physics, the view has the consequence that we cannot know which purported special-science laws are genuine laws of nature—if indeed any are. While some philosophers may accept such a view, we do not consider it a viable option. Like physics, the special sciences aim and sometimes succeed at discovering laws of nature, and any naturalistically acceptable view must respect their ability to do so.

There are two other ways that the web-of-laws theorist could try to accommodate special-science laws, though it is unclear whether either course can offer a viable account. One strategy would be to add each non-deducible special-science law to the system as an axiom. However, this choice leads back to the circularity problem, as the web-of-laws theorist has no independent way to judge which non-deducible regularities should be counted as special-science laws, as opposed to accidents. It may also be worth noting that the notion of strength gets even murkier with respect to the special sciences, since we often cannot make straightforward derivations from special-science laws.

The other possibility would be to have separate axiomatic systems for each special science. The problem with this is that it is unclear how this version of the MRL view could represent the unity of science, i.e. both the hierarchical relationships among the special sciences and the relationship that the special sciences bear to fundamental physics. The criterion that special-science laws be theorems derivable from the axioms of fundamental physics is at least able to capture a sense in which physics is fundamental, though we think it is too strong a sense. At the other end of the spectrum is the view that each science has its own disparate axiomatic system of laws. This picture is unable to reflect the fundamentality of physics at all; it seems best-coupled with Cartwright’s and Dupré’s views that science is essentially disunified. The proper account of laws in the special sciences ought to steer between these two extremes; it must reflect the unique relation that holds between fundamental physics and the special sciences without requiring anything so drastic as the reduction of the special sciences to physics.

8.3.2.3 Inductive Inference

A third major problem for the MRL view is that it cannot explain the success of inductive inference. Recall that the Humean is committed to the claim that the only restriction on possibility is logical consistency. This means that the Humean must accept that there are infinitely many worlds that look exactly like ours up to a given point and then diverge dramatically. Consider, for instance, a world in which all the regularities hold that have held so far in our world, but only until January 1, 2085. On that date, negatively charged objects stop attracting positively charged ones, salt ceases to dissolve in water, and fish fail to suffocate on land. The Humean cannot deny that this world is possible else she commit herself to some necessary connection between, e.g., negative charge and the capacity to attract positively charged things. If this world is possible, infinitely many such worlds—that follow exactly the same course as our own for a time and then diverge—are also possible. Not only are they possible, they are far more numerous than worlds in which the observed regularities are eternal. The internal observers of these divergent worlds believe, just as we do, that the regularities they have observed in the past will continue to hold in the future. But they will be wrong. The problem for the Humean is that once she is committed to the possibility of divergent worlds, she can have no reason to believe that ours is not such a world.

The difficulty that these worlds create for induction is as follows. The problem of induction is the problem of justifying the belief that the future will be like the past. It seems that we are justified in believing that the sun will rise tomorrow, that electrons will continue to repel one another next year, and that a thousand years from now it will still be the case that nothing travels faster than the speed of light. But if it is possible for a world to act just the way our does up to some future time when its electrons will fail to repel one another, then we can have no reason to believe that our world is not one such world, and so we can have no justification for the belief that electrons will always repel one another.

It is true that an external observer looking at the entire distribution of properties throughout our spacetime *would* be able to tell whether or not our world is one in which electrons always repel one another. But the problem of induction is not a problem for external observers looking at the totality of facts about a world. It is inherently a problem for observers who lack full epistemic access to the future.

One might object that *after* January 1, 2085, when electrons continue to repel one another, we can know that we are not in a divergent world where the electron-regularity stops holding on that date. Certainly this is true, but it in no way provides a satisfactory response to the problem. For there will be an infinite number of possible worlds that are identical to ours up until to some point of divergence, for every possible point of divergence—that is, for every moment in time. Some worlds will have all the same regularities that we have observed in our own up until the moments just before they end. While we can eliminate infinitely many epistemic possibilities for every point we pass in time, there will always be infinitely many more that we cannot eliminate.

The MRL account of laws is of no use in providing a Humean response to the problem of induction. Consider the only way that the view could explain how our inductive inferences are justified. The MRL view would appeal to the fact that a given regularity is an axiom or theorem in the best deductive system describing our world as the grounds for justifying our inference that the regularity will continue to hold in the future. Since Reichenbach's regularity is not found in our best deductive system, we are not justified in believing that the future will see an absence of gold chunks larger than a cubic mile. As the plutonium regularity presumably does follow from the best deductive system, we are justified in believing that there will never be a plutonium chunk larger than a cubic mile. Even if we put aside the circularity problem from the previous section, this response will not be satisfactory.

Given the possibility of the deviant worlds discussed above, as internal observers, we can never know which (set of) deductive system(s) best describes our world. Since it is both metaphysically and epistemically possible[8] that we occupy a world in which the electron-regularity has an expiration date, we cannot know what the final deductive system of our world will look like. If we are in a world in which the regularity holds eternally, then our final system will be one in which the regularity is an axiom or theorem.[9] If we are in a world in which the regularity is only temporary, the final system will not include it. We cannot know which system is the accurate description of our world until all the facts about our world are in. This entails that only an external observer looking at the totality of facts about a world can be in a position to know what the final axiomatic deductive system will be; the internal observer is never in a position to know. Since internal observers can never be in a position to know whether an observed regularity will be a part of the final deductive system, their beliefs that the regularity will persist can never be justified.[10]

Further, the Humean can offer no explanation for the success of inductive inference beyond dumb luck. This is especially troubling when we consider just how lucky it is that we have thus far found ourselves in a world where the regularities we observe continue to hold. Based on the sheer number of worlds in which these observed regularities fail to hold at some time, it is extremely improbably that we occupy a world in which they will continue to hold. Given this fact, the Humean ought to expect inductive inference to *fail*, and she cannot attribute its incredible success thus far to anything more than a cosmic miracle.

From the perspective of a scientific realist, this does not sit well. One of the driving motivations behind scientific realism is the acknowledgement that there must be a non-miraculous explanation for the success of science. The use of induction is a cornerstone of scientific theorizing and investigation. If the Humean cannot offer a non-miraculous explanation for the success of the inductive method, neither can she offer such an explanation for the success of science.

8.3.2.4 Ontological Troubles

We have shown that the so-called 'sophisticated' version of Humeanism faces several unique problems. But there is the more general problem with Humeanism that one of its commitments—namely, the commitment to quidditism about properties—is inconsistent with the view's motivations. Humeanism is motivated in part by

[8] For the Humean.

[9] Again, we are disregarding the problem of distinguishing accidental regularities that hold eternally from lawlike regularities.

[10] Since non-Humeans about laws accept natural necessity, they are not susceptible to this problem. Non-Humeans can simply deny that there are any deviant possible worlds by ascribing necessary connections to pairs or collections of properties that feature in the relevant regularities.

the empiricist attitude that we should not be committed to the existence of what we cannot possibly observe. Since, the reasoning goes, we can only ever observe the constant conjunction between events and never the actual causal relation, we ought not have a metaphysical account of causation as anything more than constant conjunction. But in denying that there is any physical or causal necessity in the world, the Humean must deny that there is any necessary connection between what a property is and what causal relations it bears to other properties. So the Humean must maintain that properties are something over and above their causal profiles, and they are thereby committed to the view that properties have quiddities.[11]

A number of philosophers (e.g. Bird [2], Hawthorne [10], Shoemaker [22]) have suggested that what physical properties *are* is determined, at least in part, by what they *do*. We can call views that fit this framework *dispositional* accounts of property-individuation. On these views, a physical property is a collection of dispositions, understood as the potential causal relations it can bear to other physical properties. Consider the property of being negatively charged. On a dispositional view, part of what it is to be this property is to have the capacity to attract positively charged things. Thus, on a dispositional view, there is a necessary connection between being negatively charged and being able to attract things that are positively charged. Since dispositional views of properties are committed to there being non-logical necessary connections in nature, they are incompatible with Humeanism. The Humean must deny that a property's causal profile plays any role in determining its nature. Thus, she must accept that there is something else that determines a property's nature, presumably an unobservable quiddity or primitive 'thisness.'

Humeanism's commitment to quidditism is deeply at odds with its empiricist motivation. Whereas the Humean rejected a metaphysically realist account of causation in order to avoid commitment to unobservables, she now finds herself committed to the existence of ineffable quiddities. Michael Esfeld [7] makes a similar point, noting that although Humean metaphysics seems ontologically parsimonious at first glance, it is not parsimonious in the end.

Not only must the Humean be committed to quiddities, she must accept that there are physically indistinguishable possible worlds—in other words, that there are differences that don't make a difference. Lewis recognized that quidditism leads to the metaphysical underdetermination of theories, which is what led him to consider the thesis of Ramseyan Humility. Quidditism entails that two worlds can share all of their physical features and yet differ only in terms of their quiddities. On the view, for instance, the role that negative charge plays in our world might be played by positive charge in another world, but the two worlds will share all of the same observable features. Since there will be multiple worlds that satisfy the Ramsey sentence describing the laws of our world, we can never know which of these worlds is our own.

Whereas the motivation for Humeanism seems to begin with a desire to keep metaphysical commitments to a minimum, it is clear that this desideratum cannot be

[11] Quiddities are to properties what haecceities are to individuals.

obtained through Humeanism. In denying the notion of causal or physical necessity, the Humean commits herself to the existence of ineffable quiddities and differences that don't make a difference—hardly a metaphysics that can be called deflationary.

8.3.3 Why Be Humean?

Even the best version of Humeanism is plagued with problems: the above arguments suggest that it cannot be formulated non-circularly, it cannot explain why inductive inference ought to be successful, and it cannot offer a satisfactory account of laws in the special sciences. So it is worth considering why exactly one might be a Humean in the first place. This question, which Tim Maudlin [17] asks in a chapter title, is one to which he finds no satisfactory answer. Maudlin characterizes Lewis's Humeanism as the conjunction of the doctrines of *Separability* and *Physical Statism*. Separability is understood as the thesis that the complete physical state of the world is determined by the intrinsic physical state of each spacetime point and the spatio-temporal relations between those points. Maudlin argues that there is no scientific motivation for Humean metaphysics and that on the contrary quantum physics strongly suggests that the world is non-separable. Consider a pair of particles in the Singlet State. If Separability held, the joint state would be determined by the intrinsic states of each of the components and their spatio-temporal relations. But since the spin of the particles is entangled, the anti-correlation between their spins in any given direction cannot be recovered from the individual spin state of each particle (the maximally mixed state). Hence, the doctrine of Separability, which Lewis's Humeanism explicitly requires, appears to be straightforwardly false.

One could, however, hold a semi-Humean view that rejects Separability but maintains Physical Statism. Physical Statism is the doctrine that all facts about a world, including modal and nomological facts, are determined by its total physical state. It has the consequence that no two worlds can have the same total physical state but differ in their laws of nature. Maudlin argues that contemporary physics also seems to give rise to counterexamples to this thesis. Consider that cosmologists regard models of General Relativity's field equations as possible ways a universe governed by GR could be. Minkowski spacetime is a model of the GR field equations (the vacuum solution) but it is not a model *only* of GR laws. One could postulate a world in which Special Relativity exhausted the facts about spacetime structure and yet some other theory of gravitation held, one which would still include vacuum Minkowski spacetime as a model. These two empty Minkowski universes would share their total physical states but differ in their laws of gravitation. Since Physical Statism, the doctrine at the heart of Humeanism, appears to be in direct conflict with some of the most basic assumptions of contemporary physics, it is unclear what could motivate one to be Humean.

Maudlin considers Hume's own motivations, which relied in part on the semantic thesis that any non-analytic claims that go beyond the realm of the empirically observable are meaningless. But hardly any of those who accept Humeanism also accept a verificationist theory of meaning, and even fewer accept Hume's epistemological view that every simple idea must come from a simple impression. 'The semantic theory that underlies Hume's own views has been thoroughly discredited,' Maudlin writes. 'Why should one have "Humean scruples" any more?' We share in his bewilderment.

8.4 Conclusion

While modality is much discussed in metaphysics and philosophy of science, little attention has been given to the question of what sort of modal metaphysics scientific realism requires. Ontic structural realists have made a reliance on physical modality part of their view of science. Indeed, Ladyman and Ross say that ontological commitment to modal structure is what distinguishes their view from structural empiricism. Van Fraassen himself claims that what ultimately drives him to reject scientific realism is the metaphysics of modality he thinks it entails. We have delineated the modal notions in science that give realism its prima facie commitment to objective modality, and we have explored the most plausible Humean account of it. Since sophisticated Humeanism cannot offer a reasonable account of special-science laws or explain the success of inductive inference, it is deeply at odds with the spirit of scientific realism. We thus conclude that the scientific realist must embrace natural necessity.

References

1. Ayer, A.J. (1956) *The Problem of Knowledge*. London: Macmillan.
2. Bird, A. (2007) *Nature's Metaphysics*. Oxford: Clarendon Press.
3. Cartwright, N. (2007) *Nature's Capacities and their Measurement*. Oxford: Clarendon Press.
4. Cohen, J. and C. Callender (2009) A Better Best System Account of Lawhood. *Philosophical Studies* **145**: 1–34.
5. Dennett, D. (1991) Real Patterns. *Journal of Philosophy* **88**(1): 27–51.
6. Earman, J. (1984) Laws of Nature: The Empiricist Challenge. In R. Bogdan (ed.), *D. M. Armstrong* (pp. 191–223). Dordrecht: D. Reidel.
7. Esfeld, M. (2009) The Modal Nature of Structures in Ontic Structural Realism. *International Studies in the Philosophy of Science* **23**(2): 179–194.
8. Goodman, N. (1983) The New Riddle of Induction. In N. Goodman (ed.), *Fact, Fiction, and Forecast* 59–83. Cambridge: Harvard University Press.
9. Hardin, C.L. and A. Rosenberg (1982) In Defence of Convergent Realism. *Philosophy of Science* **49**(4): 604–615.
10. Hawthorne, J. (2001) Causal Structuralism. *Philosophical Perspectives* **15**: 361–378.
11. Ladyman, J. (2000) What's Really Wrong with Constructive Empiricism?: Van Fraassen and the Metaphysics of Modality. *The British Journal for the Philosophy of Science* **51**: 837–856.
12. Ladyman, J. (2008) Structural Realism and the Relationship Between the Special Sciences and Physics. *Philosophy of Science* **75**(5): 744–755.
13. Ladyman, J. and D. Ross (2007) *Every Thing Must Go*. Oxford: Oxford University Press.
14. Lewis, D. (1994) Humean Supervenience Debugged. *Mind* **103**(412): 473–490.

15. Lewis, D. (2009) Ramseyan Humility. In D. Braddon-Mitchell and R. Nola (eds.), *Conceptual Analysis and Philosophical Naturalism* (pp. 203–222). Cambridge: MIT Press.
16. Loewer, B. (1996) Humean Supervenience. *Philosophical Topics* **24**: 101–126.
17. Maudlin, T. (2007) *Metaphysics Within Physics*. New York: Oxford University Press.
18. Price, H. and R. Corry (2007) *Causation, Physics, and the Constitution of Reality: Russell's Republic Revisited*. Oxford University Press.
19. Psillos, S. (2003) *Causation and Explanation*. Chesham: Acumen.
20. Ramsey, F. (1990) *Philosophical Papers*. Cambridge: Cambridge University Press.
21. Ross, D. (2009) Ontic Structural Realism and Economics. *Philosophy of Science* **75**(5): 732–743.
22. Shoemaker, S. (1998) Causal and Metaphysical Necessity. *Pacific Philosophical Quarterly* **79**: 59–77.
23. Sober, E. (1983) Equilibrium Explanation. *Philosophical Studies* **43**(2): 201–210.
24. Strawson, P.F. (1952) *Introduction to Logical Theory*. New York: Wiley.

Chapter 9
Adding Modality to Ontic Structuralism: An Exploration and Critique

Stathis Psillos

Everyone needs a little magic somewhere.
John Bigelow & Robert Pargeter [3]

Effective magic is transcendent nature.
George Eliot, Middlemarch

9.1 Introduction

Ontic Structural Realism (OSR) gives ontic priority to structures over objects. In its most extreme form (captured, admittedly, by a slogan) it states that "all that there is, is structure" [6, p. 189]. If this is true, if there is nothing but structure(s) in the world, the very idea of contrasting structure to non-structure loses any force it might have. Actually, if the slogan is right, the very idea of characterising what there is as *structure*—as opposed to anything else—becomes incoherent. Traditionally, characterising something as a structure has made full sense—and has served excellent scientific and philosophical purposes—precisely because structure was understood as an entity with slots, which could be occupied by objects and whose individuation-conditions involved objects only *qua* slot-fillers. If objects altogether go, whatever remains can be called 'structure' only if we take 'structure' to be a term of art.

Well, Ontic Structuralists are happy to 'mimic' talk of non-structure, or objects in particular, but they hasten to add that this mimicking does not imply any serious metaphysical commitment to them. Here are a couple of characteristic passages:

- The notion of objects should be reconceptualised in "purely structural terms" [11, p. 37].
- The objects play only "a heuristic role allowing for the introduction of the structures which then carry the ontological weight" [8, p. 204].

S. Psillos (✉)
Department of Philosophy and History of Science, University of Athens, University Campus, 15771 Athens, Greece
e-mail: psillos@phs.uoa.gr

E.M. Landry, D.P. Rickles (eds.), *Structural Realism*, The Western Ontario Series in Philosophy of Science 77, DOI 10.1007/978-94-007-2579-9_9,

I have criticised OS's abandonment of objects—as a distinct and separate ontic category—in my [21]. Steven French [9] has replied to this criticism. I think we have reached a stalemate. Can there be any progress in the debate?

In an attempt to pursue further my (hopefully constructive) criticism of OSR, I want to examine in some detail a key recent idea that seems to shape the very kernel of this view, viz., that structures are *modal*. Perhaps, James Ladyman is more explicit than French in requiring that structures display or possess primitive modality. He says: "the structure described by scientific theories is the modal structure of the phenomena", adding (somewhat puzzlingly) that "the phenomena have structure but they are not structure" [15, pp. 73–74]. But French too intends to give structures causal power [9].

In Section 9.2, I argue that Ontic Structuralism has to work with a notion of structure that is meant to play two roles at once: it should be abstract enough to be independent of concrete physical systems (and hence shareable by distinct physical systems) and concrete enough to be part of the causal identity of physical systems. I then reveal the tensions there are in this mixed view. In Section 9.3, I take on a more moderate version of OS—advocated by Michael Esfeld—which identifies structure with causal structure. I then argue against the resulting causal structuralist view of the world. In Sections 9.4 and 9.5, I explore a natural way to modalise structure, viz., taking structures to be structural universals. I argue that, despite all prima facie advantages, this view inherits all problems that structural universals face and in particular the so-called 'mereology or magic' dilemma. In Section 9.6, I examine a prima facie plausible way to avoid this dilemma, which is based on the claim that there are certain spatial (or arrangement) universals that capture pure structure. I explain why this view fails to offer solace to ontic structuralism. I conclude that certain plausible attempts to modalise structure leave deep scars on ontic structuralism.

9.2 Adding Modal Force to Structures

There is an immediate problem with adding modal force to structures. If by 'structure' we mean *mathematical* structure, how can it be the locus of modality? To be sure, if mathematical structures exist at all, it is plausible to think that they exist necessarily. But there is where their modal status ends. Being abstract, mathematical structures cannot enter into causal relations; they cannot support counterfactual conditionals etc. In my [21], I borrowed several conceptions of structure from mathematical structuralism (*ante rem*, *in re*) and claimed that OSR, armed with a mathematical understanding of structure, is unable to accommodate causation. Hence the modal force of structures that OS advocates is under threat.

This reading of OSR was based on French's reading of structure as primarily group-theoretic [8] and [10] as well as on (repeated) claims of the form: "[T]he structural dissolution of physical objects leads to a blurring of the line between the mathematical and the physical" [11, p. 41]. In fact, the official position is mixed. Ladyman [16, p. 24] says:

> The *ante rem* structuralism about mathematics defended by Stewart Shapiro among others, and the ontic structural realism about physics defended by Steven French and myself among others, are both metaphysical positions. They have in common the idea that relational structure is ontologically more fundamental than individual objects. There are of course important differences between them, the most essential of which is that ontic structural realism is a form of realism about the modal (causal or nomological) structure of the world, whereas *ante rem* structuralism is only concerned with mathematical reality.

French [9, p. 174] goes one step further by arguing that "the comparison with mathematical structuralism is misleading". Here is how he thinks an OS should conceive of the matter:

> The quantum structure, say, does not exist independently of any exemplifying concrete system, as in the *ante rem* case, it is the concrete system! But that is not to say that such a structure is simply *in re*, because the ontic structural realist does not—or at least should not—accept that the system, composed of objects and relations is prior to the structure. Indeed, the central claim of OSR is that it is the structure that is both (ultimately) ontically prior and concrete.

I find this kind of claim *very* puzzling. To avoid vacuity, where talk about structures is just a roundabout way to talk about actual and concrete physical systems (like a hydrogen molecule, or a water molecule, or a pair of entangled electrons or what have you), OSR should work with a notion of structure that plays two roles. On the one hand, it should be abstract enough to be independent of concrete physical systems (so that it can be said that it is shared by distinct but structurally similar physical systems; it can be represented mathematically independently of the actual details of concrete physical systems and the like). On the other hand it should be such that it should be instantiated by (and hence be part of the identity of) concrete physical systems (so that it plays a role in making a physical system what it is; it contributes to the explanation of its causal role and the like). Given these two roles (more on this below), my feeling puzzled has to do with the fact that I simply cannot see how French's claim above makes any headway in understanding how these two roles are actually fulfilled by structures as conceived by ontic structuralists. To put the point crudely, French seems to require a conception of structure which renders structures both concrete (*qua* particular spatiotemporal physical systems) and abstract (*qua* shareable by distinct physical systems). In any case, if structures are all there is, what are they said to be ontically prior to?

In my [21], I suggested that there might well be a certain understanding of OSR which does render structures modal. This is what John Hawthorne [13] has called 'Causal Structuralism'. CS is the (popular) view that properties are identified via their causal profile, that is by the causal powers they confer on their possessors. This causal profile is a network of causal relations among properties. CS is *structuralism* because it denies quidditism, viz., the view that there is something to a property—a *quiddity*—over and above its causal profile, which makes this property what it is, independently of its causal profile, if indeed it has one. On the quidditist view, two properties may have the same causal profile and yet be distinct, because they have different quiddities. Denying quidditism, we may conceive of CS as the view that properties have no intrinsic nature over and above their causal profile. So, for every property (i.e. for every non-logical or non-mathematical property), there *isn't* its causal role (profile) *and* whatever fills in (or plays) this role; there is *just* its causal role.

French is not entirely clear on CS, but he [9, p. 182] seems to be open to reading OSR as a version of CS. The official view, as it were, is that OSR can "appropriate" whatever the settled view is on whether properties are powers or not. He, then, goes on to say:

> What we are faced with is a choice between particular relations or kinds of relations having, as features, causal aspects particular to those relations or kinds and some form of underlying causal activity which imbues the relevant relations with causal powers. Granted that the former seems more clearly structuralist, I can't see why the second couldn't be incorporated as well.

This kind of move does not take us too far ahead. It is one thing to particularise causal activity to relations—denying that there is a generic causal activity underlying all relations—and it is quite another matter to endow relations with causal activities *in the first place*. The latter claim is presupposed by French in both options stated in the quotation above, and it is precisely this claim that is problematic. If relations are imbued with causal powers (idiosyncratic or generic), a story needs to be told as to how this is possible. These causal powers will either supervene on the causal powers of the properties of particulars or they will not. If they do so supervene, this move leads to causal structuralism *simpliciter*. If they do not supervene, they become mysterious: they are just posited so that the resulting relational structure has the required modal force. A story should, then, be told as to how they emerge and how they are what they are. Recall that, as French himself admits, what we are after is an account of how structures have causal powers; claiming that they actually *do* is nowhere near the required account.

Let me press this point a bit more. There are cases of relations that cannot be said to embody causal power or activity, for instance spatio-temporal relations. There are also the properties and relations of relations themselves, especially the quasi-logical ones, that cannot be imbued with causal activity, e.g., being reflexive or being asymmetric or being reducible to etc. More importantly for our purposes the (higher-order) relations that are supposed to capture the modal relations that are supposed to exist between properties and relations (e.g., metaphysical entailment, necessitation, exclusion etc.) cannot themselves be treated as embodying causal activity, on pain of circularity. Are they then identified in a non-causal way? This move would amount to attributing a kind of quiddity to them in opposition to the dicta of causal structuralism. Trying to move in between the horns of this dilemma, friends of causal structuralism (certainly Lowe and possibly Mumford) take these higher order relations to be *formal*. This characterisation might well place them in a special category *vis-à-vis* all other relations, but it is not clear at all what exactly it is attributed to these relations (what it is for them to be formal other than being second-order and non-causal) and what the independent motivation for this characterisation is.

There is perhaps a reason why French is not *so* keen on CS. Causal structuralism does not eliminate or avoid properties altogether. It dispenses with their quiddities, but, as a matter of fact, it accommodates properties and secures their existence and causal efficacy via their causal profile. Ontic structuralism would in fact require a kind of causal *hyperstructuralism* [13, p. 223], whereby causal profiles are *purely* structural as well. But then we end up with anything but a formal structure, with no modal profile at all.

What is absolutely clear is that the friends of OSR do *not* want to endorse hyperstructuralism. What is unclear is whether and how they might succeed in this. Writing on related things, Ladyman [16, p. 39] raises the following worry: "If only (...) structural aspects of the mathematical formalism of physical theories are relevant to ontology in physics, then there is nothing to distinguish physical and mathematical structure". His reply (ibid., p. 40) however is deeply puzzling:

> Physical structure exists, but what is it? What makes the world structure physical and not mathematical? Ladyman and Ross [17] advocate a kind of neo-positivism according to which when questions like this arise it is time to stop (...).

Refraining from asking a question does not eliminate the problem raised in it! One way to proceed might well be to try to see whether there can be a meaningful distinction between mathematical structure and physical structure that can be raised *within* OSR. It will turn out that the problems faced by attempts to draw such a distinction are bigger than the possible benefits of drawing it. But it is worth exploring the options, before we pass a judgement.

9.3 Causal (Hypo)structuralism

There is, in the market, a moderate version of OSR, advocated by Michael Esfeld among others, according to which "physical structures are networks of concrete, qualitative physical relations among objects that are nothing but what stands in these relations, that is, do not possess an intrinsic identity over and above the relations in which they stand" [7, p. 180]. M-OSR, let's call it, does not do away with objects altogether. One (certainly, I) may have qualms about what exactly it is for something to have 'no intrinsic identity over and above the relations in which it stands'. I take it that this can only tell us *how many* objects there are, without saying a lot more about *what* they are. But let's pass that over in silence. The key claim of interest is that M-OSR adopts Causal Structuralism and thereby promises to ground/explain the modal features of ontic structuralism. Indeed, Esfeld advertises his programme as filling a lacuna (this concerned with modality) in OSR [7, p. 180].

Before we discuss how the lacuna is filled and how successfully this is performed, let me raise a couple of preliminary points. The master argument for Causal Structuralism is anti-quidditism. More specifically, the standard rendition of the master argument is that if properties have quiddities, these will end up being unknowable. Indeed, Lewis has famously called this view 'Ramseyan Humility'. Lewis is happy with the humility—hence, there is no problem with positing quiddities—but others think that the cocktail Quidditism & Humility is poisonous. Hence, they deny Quidditism, which implies Humility. I am not going to review this debate here, in the interest of speed. Suppose, however, that quidditism is wrong (though I very much doubt it). Esfeld (and others) think that Causal Structuralism avoids Quidditism. In a sense, it does, since if all properties are powers, and if powers are individuated by their causal profile only, there is no further issue of what makes a causal power what it is. Nor is there room for positing an extra individuating factor which marks the identity of the power independently of its causal profile. Two putatively distinct causal powers which have exactly the same causal profile are one and the same property. Esfeld (and others) also think that Causal Structuralism avoids Humility. In a sense, it does, since if all properties are powers, and if powers are the kind of entities that cause things to happen, and if knowledge requires causal contact with the thing known, knowledge of properties is *in principle* possible.

There is a certain sense, however, in which Causal Structuralism simply relocates the quiddity. The identifying feature of a property is simply transformed from a local individuating feature to a global feature of the causal network in which the property participates. What is more, this global feature is no longer individuating! Let me explain.

Causal Structuralism advances a holistic account of the individuation of properties. Strictly speaking, this is not necessary. There can be an 'essentialist' version of CS according to which not all elements that are parts of a property's causal profile are essential to this property being what it is. On an essentialist causal structuralism, only some parts of the causal profile of a property P (perhaps some particular relations to some other properties or some particular effects) fix the identity of P. Though this *is* a genuine option, it is hard to defend it unless there is a natural distinction to be drawn between the essential and the accidental parts of a causal profile.

To the best of my knowledge, most causal structuralists are in favour of an anti-essentialist holistic individuation of causal powers. In its clearest form, this position is found in [19], but is also explicitly present in [5] and others. Mumford says that "a property's identity is fixed by the (causal) role it plays in relation to other properties" but adds that though the identity of a property "is fixed by relations to other properties, its existence has no ontological dependence on those properties [19, p. 171]. Later on, he explains that he accepts holism, whereby "the world is a single whole, composed of properties whose essence and identity are determined by their place in that whole" [19, p. 184]. And again: "the properties that are real in a world must (...) form an interconnected web: a system with no property standing alone or outside".

If this image is taken seriously, a property cannot be identified, unless what all other properties to which it is related are has already been specified; that is, unless all other properties have already been identified. But since this tangle arises for *any* property whatever, it follows that no property can be identified unless some other properties have already been identified, and because of this, no property can be identified *simpliciter*. All we get, at best, is a web of causal profiles, but no other way to tell how the several parts of the web are related to (or flow from) certain properties. This way to understand the identity of properties was motivated, at least partly, by an attempt to avoid the supposedly mysterious quiddities *qua* unknowable metaphysical identifiers of properties. Nothing much is gained by replacing them with a more mysterious holistic network of relations among properties, which is supposed to confer identity on properties, without in the end identifying any of them. Quiddities are not dispensed with; they become a global *totalitas*.

Hawthorne discusses a version of this problem and notes that it is not too disturbing. His idea is this. Take all the laws that characterise all properties in the world and express them in a lawbook. Then use the Ramsey-Lewis technique to Ramsify away the properties, by replacing each property-name by a distinct variable and prefixing the resulting open sentence with an equal number of existential quantifiers. Call this, Hawthorne says "the Ramsified lawbook". He then [13, p. 220] adds:

> We can now articulate causal structuralism very easily, and whatever its merits, we cannot be accused of vicious circularity. Since the variable 'F1' replaced [property name] A, we can give a theory of the individual essence of A by the open sentence you get by dropping the existential quantifier prefixing 'F1'. According to causal structuralism, it is a necessary truth that anything that satisfies that open sentence is identical to A. Generalizing, the causal structuralist will say that any natural property can be defined by a suitable open sentence delivered by the Ramsified lawbook for that property.

Fair enough! But this strategy won't take us very far if *all* properties are taken to be structurally identified powers that are Ramsified away. For if all properties are identified by their relations to all other properties and all properties are Ramsified away, nothing will be left to tell us what these properties are. The suitable open sentence delivered by the Ramsified lawbook for a certain existentialised away property will include all other existentialised away properties; hence it will not specify any of them. All it will succeed in identifying is the whole network of properties that satisfy the Ramsified lawbook, without identifying any of them in particular. Here again, we get, at best, a *totalitas* (the Ramsified lawbook) and a specification of properties in relation to it. But if everything is Ramsified, even this relative specification will leave us in the dark as to what property is what. There is a way out, of course, and this is to keep some part of the lawbook unRamsified. But this would imply that at least some properties get their identity in a different manner.

Mutatis mutandis, the same goes for Alexander Bird's attempt to disarm the problem with holistic individuation noted above. He favours a graph-theoretic account of the relations among causal powers (or potencies, as he prefers to call them). The details need not detain us here. The relevant point is that according to Bird if the relations that structure the fundamental properties have certain features or characteristics—they are asymmetric, non-irreflexive relations—then the properties so structured can be individuated. As he [4, p. 142] put it:

> Thus if we consider that the fundamental properties are structured by the asymmetric, non-irreflexive relation between a power and its essential manifestation property, then we can see that there could be any number of fundamental properties, represented by the vertices on directed graphs that may contain loops.

There is no disagreement with what Bird asserts. However, the nodes in the holistic causal network are told apart from each other because (and only when) the network is of a certain sort: the relations that structure it have certain properties which are individuated non-structurally and independently of their causal role (though it is doubtful that they have a causal role in the first place). To put the point a bit provocatively, causal structuralism (of the sort discussed so far) does offer individuation conditions for causal powers that acquire their identity by their place in a network of causal profiles, provided that *some* properties or relations (or, indeed, the network itself), get their identity independently of the their place in this network. Actually, for causal structuralism to get off the ground, causation itself must be a relation which is identified independently of its role in a causal network. But this is a different story. The bottom line is that Causal Structuralism ends up being causal *hypostructuralism*.

Mumford [19, pp. 186–187] appeals to a standard move, which is common to all friends of powers. This is that some powers are, ultimately, identified by the effects they have on us and our sensory modalities in particular. He [19, p. 187] says:

> We are able to interact with properties. Among the effects they have in their cluster of causal powers, are the effects they have on us, namely their phenomenal appearance. We can thus know properties either by the phenomenal appearance they cause in us or by the phenomenal appearance on us of other effects they cause. Some of the relations borne by properties are thus experienced and in this way we are able to break into the circle of interdefinability for the nature of a property.

In a similar fashion, Chakravartty [5, p. 136] says:

> Every case of warranted causal property attribution is facilitated by some properties that are known independently of a knowledge of their further effects. These latter property instances are the direct objects of our perceptions.

There is no doubt that *some* properties have effects on us. But if we took a property's effects on us to give us privileged access to the identity of a property—assuming we can tell which property has what effects on us—the very idea that a property is identified by its relations to all other properties to which it is related would be threatened. As noted already, some such relations would become the essential identifiers. We would therefore end up with essentialist causal structuralism. To sum up my first preliminary point, there is a sense in which Causal structuralism replaces *quiddity* with *totalitas*. And there is also a sense in which this is avoided only by retreating to causal *hypo*structuralism.

My second preliminary point is that there is also a sense in which CS replaces Humility with Audacity *only in name*. The friends of CS take pride in claiming that if properties are powers, they are in principle knowable; hence Humility is avoided. Esfeld sums up this sentiment by saying that on CS "what the properties are can in principle be discovered via the effects they produce" [7, p. 184]. If CS holds sway on all properties, Humility (associated with quidditism) is replaced by Audacity: *all* properties can be known. Even without a lot of reflection, this claim appears too strong. The chains by means of which causal properties are detected (and hence known) are long, complicated and sometimes devious. Some properties—too remote causally, or too shielded by other causal properties—might not be known, even if CS is true. On reflection, however, things are worse. Given the problem identified above concerning the individuation of properties, it might well turn out that even if the total network of causal profiles—what I called *totalitas*—might be knowable, what properties play what role *within* this totality might not be knowable after all—unless the nexus of interdependent and interconnected properties is broken at several places (e.g., at the level of phenomenal effects on humans) in such a way that among the several effects that a property has, the effect that has on us is singled out as *the* identifier of this property. Not only is CS in danger of being abandoned. More importantly, humility is still with us, despite claims to the opposite, for all those properties that are not fortunate enough to yield effects on us.

9.4 The Abstract and the Causal

These, I am afraid, are preliminary skirmishes. The key battle is still to come. At stake is Esfeld's claim that CS can make good on the OS promise to modalise structure. He [7, p. 185] declares: "the fundamental physical structures are causal in themselves so that there is no need to postulate underlying causal properties". And he (ibid., p. 187) adds:

> if the fundamental physical structures are modal structures, being the power to produce certain effects, then (...) any difference in the fundamental structures, accounting for there being two different types of arrangements of fundamental structures in two possible worlds, automatically leads to some difference in the effects that these structures produce and thereby also to some difference in the domain of observable phenomena.

The thought here is that different structures produce different effects and in particular different observable effects. This, however, does not seem quite right. Take Newtonian mechanics, where $\mathbf{F} = m\mathbf{a}$, and compare it with a reformulation of it, according to which $\mathbf{F}$ is always the vector sum of two more basic forces $\mathbf{F}_1$ and $\mathbf{F}_2$. Suppose further that $\mathbf{F}_1$ and $\mathbf{F}_2$ are such that they sustain each other and can act only in tandem to produce acceleration. Suppose further that $\mathbf{F}_1$ and $\mathbf{F}_2$ have no other effects. We have two modally-laden structures which are non-isomorphic but, nonetheless, empirically equivalent.

Indeed, it is only by fiat that CS can block the following from being a genuine possibility. Two properties A and B act in tandem to generate a certain causal profile Q. Suppose, further, that A or B, taken individually, do not have any further causal role. Causal structuralism entails that, all else being equal, a world W_1 with A&B having causal profile Q would be identical with a world W_2 in which a *single* property C has causal profile Q. We may never be able to figure out whether we live in W_1 or W_2, but to make sense of this metaphysical difference we need to go beyond causal roles.

Still, the main thought remains: CS can make OSR more attractive by taking physical structures to be genuinely causal, their essence being their power to produce certain effects. Esfleld takes it that one of the advantages of this move is that the distinction between mathematical and physical structure is no longer blurred. In what follows, I shall argue that Esfeld's blending of ontic structuralism with causal structuralism is misplaced.

Concrete structures are best seen as relational systems—that is systems of entities having properties and standing in certain definite relations to each other. As such, they are concrete systems, located in space and time. They can stand in causal relations to other systems, where, as a rule, these causal relations are determined, at least partly, by the properties and relations of the elements of the relational system. They have a structure in the sense that they have a certain spatial-geometric arrangement. Their unity—*qua* concrete relational systems—is causal-nomological.

Qua concrete structures, relational systems can share structure; they can instantiate a common structure. In fact, two or more distinct relational systems fall under the same *type* partly because they share structure. Two or more water molecules—*qua* concrete relational systems—are *water* molecules precisely because they have the structure of a water molecule, which is a type of structure distinct from other types of structure not just on the basis of the elements that compose it but also on the basis of their structural properties. The structure or form of a water molecule is an abstract entity. It is shareable among distinct (and spatially separated) particulars. Unlike a concrete water molecule, it has slots—which can be occupied by distinct elements. Structure, in general, is like a universal which is instantiated in many particulars. It is an one over the many; a recurring and repeatable characteristic of distinct particulars.

The question then, as already noted in Section 9.2, is: how can this structure—*qua abstract*—be modal? How can it have modal features? Can it stand in causal relations? Can it support counterfactuals? If we think of structures as universals, that is properties, it transpires that they can be both abstract and modal. Properties, *qua* universal, are abstract—they are not concrete; they are shareable by many particulars; they are not 'in' space and time in the way particulars are—and yet they are causal in that they can and do cause things to happen. They can also stand in nomological relations, relations of counterfactual dependence and the like. At least this is what a lot of realists about properties *qua* universals think. But what kind of universals could structures be?

9.5 Structural Universals to the Rescue?

The most natural suggestion is that *qua* universals, structures should be taken to be *structural universals*. Structural universals have been explicitly introduced in order to account for the sharing of structure among particulars. They have been seen as universals of structure. Bigelow and Pargeter [3, p. 82], for instance, say:

> Chemical compounds are structures which are formed from the elements. The property of *having* such a structure is a universal which is related in quite distinctive ways to the universals which determine the elements. It is a *structural* universal.

And David Lewis [18, p. 82], while opposing the idea of structural universals, admits that a good reason for accepting structural universals is to explain "structural resemblance as the sharing of universals". The key idea is that the fact that distinct particulars are composed of similar parts which are arranged in a similar way—that is the very idea of sharing structure—might be explained by positing structural universals.

Structural universals are *universals*—they are repeatable and recurring features of the world; they are instantiated by spatio-temporaly distinct particulars; they are in some non-spatial sense 'in', or 'part of', the particulars that instantiate them. Structural universals are *structural*: they have other universals as parts (again in non-spatio-temporal sense) and the particulars that instantiate them have proper spatio-temporal parts in which the universals that are 'parts' of the structural universal are instantiated. *Methane* is a standard example. A methane molecule is made of one carbon atom and four hydrogen atoms, arranged in a certain spatial way. The bonds between the carbon atom and the hydrogen atoms are co-valent. *Methane* molecules—actual particulars—are supposed to instantiate the methane universal. This is a structural universal in that its components (in a non-spatio-temporal sense) are two monadic universals (being carbon, being hydrogen) and a dyadic universal (being bonded). Actually, within a concrete particular structure which is a methane molecule (by virtue of instantiating the universal *Methane*), the universal *Hydrogen* is instantiated four times, the universal *Carbon* is instantiated once, and the universal *Bonded* is instantiated four times. We will come to the difficulties that this generates in a moment, but for the time being let us explore the idea that *Methane*—*qua* structural universal—is both abstract *and* modally laden. There is a pattern of entailments such that, for instance, when the universal *Methane* is instantiated, the universal *Hydrogen* is instantiated too. There is also a pattern of exclusions such that, for instance, when the universal *Carbon* is instantiated as part of *Methane*, the universal *polar bond* is not instantiated. Besides, the very idea of structural universals, allows for the possibility that there is 'structure' all the way down; that is, that there are no simple universals at all.

This kind of account can capture Esfeld's view that physical structures are causal. It's *not* part of the theory of structural universals that properties are powers, but there is no incompatibility here at all. So coupled with causal structuralism, structural universals can account for "the essence of a causal structure" being "in the power to produce certain effects [7, p. 188] while at the same time ground the obvious fact that causal structures—*qua* structures—are shareable. Esfeld, to be sure, talks of *fundamental* physical structures and focuses his attention on quantum structures of entanglement. But I take it that this is a side issue. There is no principled problem in applying the theory of structural universals to quantum systems, and conversely staying at the level of molecular structures does not detract from the fact that there is a way to accommodate modal features to structures by going for structural universals.

An advantage of going for structural universals is that there might be a way to explain sameness of structure in terms of isomorphism. Particular concrete systems can be said to be isomorphic to the structural universal they instantiate. Besides, particular concrete systems can share the same structure by instantiating the same structural universal. This can be explained, as Armstrong does it, by a process of abstraction. We start from concrete physical systems, e.g., methane molecules and proceed by abstracting away the particulars. What thereby remains is a pattern of interrelated universals. This is a structural universal which can be described, as Armstrong [2, p. 432] put it, as "an individual that is a carbon atom, four further individuals that are hydrogen atoms, and where ... etc. etc.".

This cannot be quite right. The universals that constitute the structural universal occur once in it. The universal *Hydrogen*, for instance, is one; it might be instantiated four times in the methane molecule, but this does not mean that it occurs four times in the *Methane* universal—if it did, universal *Hydrogen* would not be a proper universal. Lewis [18], who identified this problem first, noted that structural universals defy mereology. He then presented the friends of structural universals with a dilemma: either structural universals have literally other universals as proper *parts*, but then they cannot be isomorphic to their instances, or structural universal are mereologically atomic but then it becomes magical how they share structure with the particulars that instantiate them; how, in particular, they impose a certain structure on the proper parts of the particulars that instantiate them.

There are various ways in which the friends of structural universals have replied to this dilemma, but the bottom line is to claim that structural universals have a *sui generis* non-mereological constitution. Armstrong captured this, at least partly, by denying that structural universals have parts—as opposed to *constituents*. He also toyed with the idea that the non-relational constituents of a structural universal are *particularing* universals, that is they are such that we can speak of them as having universals as instances—for instance, *Hydrogen* is a particularising universal in that it can have four hydrogen universals as instances in the structural universal *Methane* [1, p. 88]. Lewis called "amphibians" these particularised denizens of structural universals and claimed, quite correctly I think, that positing them makes things a lot more complicated. For instance, how many universals of *Hydrogen* do we now have? One? One plus all the particularised instances?

Bigelow and Pargeter, on the other hand, argued that a structural universal *R* is a relational property of a particular, where the relational property is such that it stands in "a pattern of internal relations of proportion to other properties [3, p. 88]. Accordingly, the structural universal *Methane*

> relates the molecule to various properties. These properties are being carbon, being hydrogen, being bonded. Being methane, then, is to be identified with a highly conjunctive second-order relational property of an individual (molecule): the property of having a part which has the property of being hydrogen, and having a part which is distinct from the first part which has the property of being hydrogen, and.. [3, p. 87].

Still, this is supposed to be a non-mereological mode of composition, which is characterised by a pattern of essential internal relations among properties and relations. Part of the problem with this move has to do with the appeal to essentialism. It should be accepted as a primitive (modal) fact that there is an essential relation between *being methane* and *being carbon*. This is already magical enough. But as Katherine Hawley [12] has noted, even if this were granted, it does not follow from it that there is a link between the thus understood structural universal *being methane* and the patterns of co-instantiation of the universals *Carbon*, *Hydrogen* and *Bonded* that characterises a *Methane* molecule. In other worlds, that *Methane* essentially involves *Carbon*, *Hydrogen* and *Bonded* related by internal relation *R* does not, as it stands, explain why a particular methane molecule has the structure it does.

The problem that Lewis has identified is that there should be a nontrivial explanation of how the structural universal shares structure with the particular it is instantiated in. This problem becomes more acute when we consider cases in which two structural universals which are 'made of' the same universals are structurally distinct. The standard example is butane and isobutane. Butane molecules are made up of four carbon atoms, ten hydrogen atoms and thirteen co-valent bonds in a particular configuration. Isobutane (methylpropane) molecules consist of exactly the same atoms as butane but in a different configuration. *Butane* and *Isobutane* have the same components (the simple universals *carbon*, *hydrogen* and *bonded*); the same number of instances of these universals; and yet they differ in structure because their components are combined in different ways. Their molecular diagrams are given in Figs. 9.1 and 9.2.

Cases such as these suggest that distinct structural universals can be composed of exactly the same parts and this defies the principle that the parts determine the whole (and in particular the same parts-same whole principle). This is known as the Principle of Uniqueness of Composition (PUC): given some parts, there is only one whole they can compose. Lewis adheres to this principle and hence denies structural universals. Armstrong, on the other hand, accepts structural universals and defies PUC arguing that states of affairs violate it anyway. [Take a non-symmetrical relation *R* and two particulars *a* and *b*. PUC suggests that there is only one mereological sum with these three as parts, but state of affairs *Rab* is different from state of affairs *Rba*. So states of affairs violate PUC.]

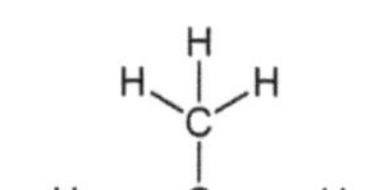

Fig. 9.1 Molecular diagram of Butane

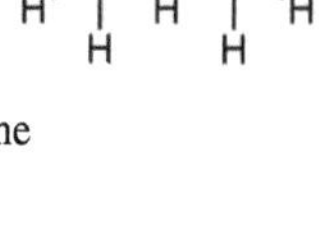

Fig. 9.2 Molecular diagram of Isobutane

9.6 Arrangement Universals

But there is a different way to proceed, which is relevant to our current concerns. This has been explored by Javier Kalhar [14]. If you think of it, *Butane* and *Isobutane* are different because they instantiate different spatial arrangements, or better, different bonding arrangements. This is obvious by the molecular diagrams above. Kalhat's idea is that *Butane* and *Isobutane* are different structural universals because they involve—as parts—different arrangement universals, alongside the universals *Carbon*, *Hydrogen* and *Bonded*. These arrangement universals are *being butane-like structured* and *being isobutane-like structured.* That *being butane-like structured* is a universal is obvious since it can be instantiated—*qua* spatial arrangement—not just by carbon atoms and the like but by anything whatever. (Similarly for *being isobutane-like structured.*) But *being butane-like structured* is also shared by all butane molecules. It explains their structural similarity and also the difference between butane molecules and isobutene molecules. Arrangement universals are, to be sure, second-order universals; more specifically, second order relations over first order relational (*Bonded*) and non-relational (*Carbon, Hydrogen*) universals. But this is not a problem that needs to give us pause. What's interesting is that these arrangement universals can be seen as the product of a double abstraction. First, the particulars are abstracted away (and we get the structural universal, *à la* Armstrong); second, the first-order universals are abstracted away, and we get a spatial structure, viz., the *pure* structure of the structural universal. So the arrangement universal *being butane-like structured* could be represented as follows:

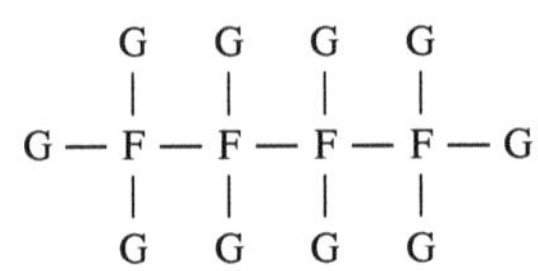

If arrangement universals are parts of structural universals, it follows that PUC above need not be violated. Structural universals can be distinct because they have as parts distinct arrangement universals. In this sense, the structure of the structural universal is part of its very constitution; it contributes to making it what it is and to what modal features it has. This account seems to suit particularly well Esfeld's approach, according to which physical structure is modally laden. It also helps explaining how "different types of arrangements of fundamental structures" lead to some difference in the effects they have [7, p. 187].

Admitting spatial universals is a step forward in this debate at least in the sense that we can now think of the *structure* of a structural universal as something repeatable and shareable. It is instantiated by the particular that instantiates the structural universal, but it is also instantiated by other particulars. It is a genuine one over the many. Besides, it can be instantiated by distinct types of particulars, making sense of the claim that, for instance, a methane molecule and a toy-model arrangement with spheres and pegs can share structure (thereby explaining how the toy-model can represent the structure of the methane molecule). More importantly, being a universal, it can embody modality; it can enter into causal or nomological relations etc.

We therefore seem to have a conception of physical structure (*qua* structural universal) which is both abstract (repeatable, recurring, shareable) *and* modal. But there is bad news too: we are not yet done with the mereology vs magic problem. Spatial universals—structures—have slots alright, but they are 'filled in' by other universals many times over. This is not quite right, since the universal *Hydrogen*, for instance, is one and not the 10 particularising instances or whatever required to 'fill in' the butane-like structure. *Butane* and *isobutane* do differ because they have different structural parts (spatial universals), but we still do not get isomorphism with particular molecules, since a particular butane molecule, for instance, has 10 hydrogen atoms, whereas the universal *Butane*, alongside the *being butane-like structured* universal, has one *Hydrogen* as its part (since *Hydrogen*, *qua* universal, is one and not ten).

It follows that an appeal to spatial universals cannot ultimately offer a mereological account of the relations between the structural universal and the particular that instantiates it. It can certainly explain why PUC need not be violated if two or more universals have some of their parts the same, provided they differ in their structure (in their arrangement universals). But it fails to explain the relation between the structural universal and the particular in which it is instantiated as one of isomorphism.

There is a further problem with spatial universals in particular, which would remain even if all other problems disappeared. This is that there is no clear sense in which they can be seen as powers or as embodying power. If anything, arrangement (spatial) universals should be seen as categorical properties, capturing structural arrangements among universals. What is more, a spatial universal (a certain geometrical arrangement, let us say, *qua* structural universal) is not a physical structure in the sense that a concrete molecule is. It could be isomorphic to a concrete physical structure, if all of the above problems were indeed resolved, but the isomorphism would not hold between two physical structures, but rather between a concrete physical structure (e.g., a concrete molecule) and a structure such that it would much more plausible to think of it as a geometrical structure.

Trying to disarm Lewis's criticism of structural universals, Hawley [12] has recently claimed that the dilemma 'mereology or magic' is false. She explored a different way forward, arguing that there is space for "a non composition relation" of the structural universal by its constituents. The details of her way forward are interesting, exploring the possibility of viewing composition as partial identity. But when it comes to the crunch (how does a structural universal impose a certain structure on the parts of the particulars that instantiate it?), what she says is rather puzzling. She says:

> Perhaps there is sense in which *being methane*, *being butane*, and being isobutene stand in a different relations to the same parts (*being carbon*, *being hydrogen* and a bonding relation). What relations? Well, those relations that underpin the relevant patterns of co-instantiation.

She does admit that this is not terribly illuminating, but adds that for friends of structural universals the difference between *Butane* and *Isobutene* is not brute; they can see this difference "as grounded in the different relations each universal bears to its parts" [12, p. 129].

But why should that be so? Why should it be the case that the relations that "underpin the relevant patterns of co-instantiation" are exactly the relations that structure the structural universal? Without further explanation, it seems we are being asked to accept this in the spirit of natural piety.

There have been indeed other attempts to characterise structural universals. Armstrong, in his later work [2], has characterised them as *types* of states of affairs. Pagès [20] has criticised this view and has gone for an account of structural universals in terms of formal relations among first-order properties and relations. But Pagès's account seems, in the end, to replace structural universals with structures of universals [2, p. 432], the idea being that the unity of the structural universal is lost.

9.7 Concluding Thoughts

Structural universals, combined with the claim that properties are causal powers, were meant to offer a way to explain how physical structures have modal force while at the same time are abstract and shareable among the particulars that instantiate them. This kind of avenue had not been explored so far. But despite its initial promise, it stumbles over important problems that structural universals face. The key problem is that though structure is meant to represent by isomorphism, structural universals fail to do that, despite some ingenious attempts to make them succeed. This has a direct bearing on the modest version of ontic structuralism, which aligns ontic structuralism with causal structuralism and aims to pin modality on physical structure. I am not claiming there are no other ways to think of physical structure. What I *am* claiming, however, is that if we take talk of physical structure *seriously*—if that is, we think of structure as a universal of a sort, recurring and repeatable and being instantiated by different concrete relational systems—thinking of it along the lines of structural universals is both natural and initially promising. If the arguments above hold any water, the promise is not fulfilled.

There is a certain optimism around that causal structuralism is the right way to think of properties. Esfeld [7, p. 192] sums it up thus:

> The metaphysics of causal properties holds hence all the way down from common sense including the experience of ourselves as agents in the world via the special sciences to fundamental physics. It therefore provides for a complete and coherent view of the world that reaches from fundamental physics via biology to psychology and to the social sciences. The argument for the metaphysics of causal properties, taking, as physics teaches us, the form of a metaphysics of causal structures, cannot simply be that it is anchored in common sense. The argument is that it leads to a complete and coherent view of the world, including all the domains of empirical science, and avoiding a gap between metaphysics and epistemology by not having to postulate that there is something in the world whose essence is a pure quality that can in principle not be known because it does not make any difference.

If what was said above has any grain of truth, there are important cracks in the causal structuralist 'complete and coherent view of the world'. Indeed, there are cracks in the structuralist metaphysics anyway.

Acknowledgments Versions of this paper were presented at the SIFA conference in Padova in September 2010 and at a seminar in the University of Cologne in January 2011. Many thanks to the participants for many useful comments and in particular to Mauro Dorato, Steven French and Andreas Hüttemann.

References

1. Armstong, D.M. (1986) In Defence of Structural Universals. *Australasian Journal of Philosophy* **64**: 85–88.
2. ———— (1993) A World of States of Affairs. *Philosophical Perspectives* **7**: 429–440.
3. Bigelow, J. and R. Pargeter (1990) *Science and Necessity*. Cambridge: Cambridge University Press.
4. Bird, A. (2007) *Nature's Metaphysics*. Oxford: Oxford University Press.
5. Chakravartty, A. (2007) *A Metaphysics for Scientific Realism*. Cambridge: Cambridge University Press.
6. Da Costa, N.C.A. and S. French (2003) *Science and Partial Truth*. New York: Oxford University Press.
7. Esfeld, M. (2009) The Modal Nature of Structures in Ontic Structural Realism. *International Studies in the Philosophy of Science* **23**: 179–194.
8. French, S. (1999) Models and Mathematics in Physics: The Role of Group Theory. In J. Butterfield and C. Pagonis (eds.), *From Physics to Philosophy* (pp. 187–207). Cambridge: Cambridge University Press.
9. ———— (2006) Structure as a Weapon of the Realist. *Proceedings of the Aristotelian Society* **106**: 167–185.
10. ———— (2011) Metaphysical Underdetermination: Why Worry? *Synthèse* **180**(2): 205–221.
11. ———— and J. Ladyman (2003) Remodelling Structural Realism: Quantum Physics and the Metaphysics of Structure. *Synthèse* **136**: 31–65.
12. Hawley, K. (2010) Mereology, Modality and Magic. *Australasian Journal of Philosophy* **88**: 117–133.
13. Hawthorne, J. (2006) *Metaphysical Essays*. Oxford: Oxford University Press.
14. Kalhat, J. (2008) Structural Universal and the Principle of Uniqueness of Composition. *Grazer Philosophische Studien* **76**: 57–77.
15. Ladyman, J. (2001) Science, Metaphysics and Structural Realism. *Philosophica* **67**: 57–76.
16. ———— (2007) Scientific Structuralism: On the Identity and Diversity of Objects in a Structure. *Aristotelian Society Supplementary Volume* **81**: 23–43.
17. ———— and D. Ross (2007) *Every Thing Must Go*. Oxford: Oxford University Press.
18. Lewis, D.K. (1999) *Papers in Metaphysics and Epistemology*. Cambridge: Cambridge University Press.
19. Mumford, S. (2004) *Laws in Nature*. London: Routledge.
20. Pagès, J. (2002) Structural Universals and Formal Relations. *Synthèse* **131**: 215–221.
21. Psillos, S. (2006) The Structure, the Whole Structure and Nothing but the Structure? *Philosophy of Science* **73**: 560–570.

Chapter 10
Ontological Priority: The Conceptual Basis of Non-eliminative, Ontic Structural Realism

Anjan Chakravartty

> *...she looked up, and there was the Cat again, sitting on the branch of a tree... this time it vanished quite slowly, beginning with the end of the tail, and ending with the grin, which remained some time after the rest of it had gone.*
> *"Well, I've often seen a cat without a grin," thought Alice; "but a grin without a cat! It's the most curious thing I ever saw in my life!"*
> *— Lewis Carroll, Alice's Adventures in Wonderland*

10.1 Clarifications Ab Initio

Realisms and antirealisms in the context of discussions of scientific knowledge have had a knack for reinventing themselves, and this is potentially a good thing. Though some despair at the prospect of seemingly perennial debates, the glass half full is that new insights are often provoked by means of these novel formulations. In this paper I consider a recent formulation of scientific realism, the core of which amounts to a provocative metaphysical doctrine. The family of views to which this innovation belongs is called "structural realism" (SR); the relevant genus within this family is now commonly referred to as "ontic structural realism" (OSR); and the novel species under consideration here is something that I will call "non-eliminative OSR", to contrast it with its older and more widely problematized sibling species, eliminative OSR. I will argue that the core metaphysical doctrine underlying non-eliminative OSR, advocating an "ontological priority" of the relations of objects and properties over the objects and properties themselves, is no less problematic. The result is a dilemma for those who would subscribe to OSR in either its eliminative or non-eliminative forms, in hopes of finding a promising way forward for realism in the context of scientific knowledge.

A. Chakravartty (✉)
Department of Philosophy, University of Notre Dame, 301 Malloy Hall, Notre Dame, IN 46556-4619, USA
e-mail: chakravartty.1@nd.edu

E.M. Landry, D.P. Rickles (eds.), *Structural Realism*, The Western Ontario Series in Philosophy of Science 77, DOI 10.1007/978-94-007-2579-9_10,

I will begin with a brief sketch of the idea of OSR, in order to foreground the metaphysical issues under consideration here. SR is the view that insofar as scientific theories offer approximately true descriptions of things in the world, they do not tell us about the underlying nature of reality—that is, about the qualitative natures of things underlying observable phenomena. Rather, they tell us about the *structures* of things.

The position comes in two broad flavours: epistemic (ESR); and ontic. ESR places an epistemic restriction on scientific knowledge in response to sceptical concerns arising from the history of theory change in the sciences. It holds that we can know structural aspects of unobservable parts of the world, but nothing about the natures of those things whose relations define these structures in the first place. The history of scientific change has surely taught us, so the story goes, that the objects themselves and their first-order properties are simply beyond our grasp, but that structural knowledge is likely to be preserved in some form over time. OSR, more radically, began as the view that at best we have knowledge of structural aspects of the world, because there is in fact nothing else to know. The idea that there are objects, for example, conceived of as things that stand in relations to one another, is according to this view a vestige of an outdated metaphysics. The motivation for the revisionary metaphysics of eliminative OSR has its source in modern physics. Quantum mechanics, for example, appears to underdetermine the nature of subatomic particles with respect to the question of whether they are individuals or not, thus generating the possible worry that our best physics does not yield any definitive picture of their ontological status. As James Ladyman [17, p. 420] put it: "it is an *ersatz* form of realism that recommends belief in the existence of entities that have such ambiguous metaphysical status".[1]

That was then. Since the initial formulation of OSR in these eliminative terms, the position has been the subject of serious scrutiny and a number of challenges, commonly premised on variations of the following observation: given the way "structure" is usually understood, in terms of relations between certain relata, it seems peculiar to say that there are structures in world, but nothing that is structured per se. The intelligibility of the position thus remains, in the eyes of many critics, a promissory note at best; the promise is to make intelligible the notion of concrete, causally efficacious things made up of nothing but structure. What if it were possible, however, to reconceive OSR in less radical terms, avoiding the commitment to the seemingly paradoxical notion of concrete relations in the absence of relata? It is this prospect, the prospect of a non-eliminative OSR, that several advocates of the position have put forward more recently, not least as a means of evading the recurring metaphysical worry facing its original formulation. It is this new non-eliminative position that is the primary subject of the following discussion.

[1] In the contemporary debate, ESR is canonically associated with Worrall (e.g. [31], and OSR with French [13] and Ladyman [17]). Both positions have attracted a variety of adherents and critics since, however. For a comprehensive summary of the literature, see [18].

Before delving into the details, it will prove useful to clarify some of the central concepts on which this discussion will rely. At the heart of OSR there is one thing on which all proponents of different stripes can agree: an emphasis on structures or structural relations at the expense of things putatively standing in those relations—their relata. This broad agreement, however, masks a lack of clarity regarding the central issue of how key terms such as "structure", "relation", and "relata" are to be understood, so let me clarify how these terms will be used here. Firstly, what of the term "structure"? Consider a set of elements and various relations defined over them. We might identify structure with the higher-order logical or mathematical properties of those relations, as we do when we say that the structure of the set is shared by others that are isomorphic to it. This is a relatively abstract conception of structure. On the other hand, we might identify structure with the specific relations between the elements themselves, which is to think of structure less abstractly and more concretely. The concept of *concrete* structure, as I intend it here, applies to first-order relations between specific kinds of relata. For example, take an equation describing relations between the magnitudes of certain properties, such as the ideal gas law. Here we have a representation of concrete structures, *viz.* first-order relations between specific kinds of relata, the properties of pressure, volume, and temperature. By "structure" I will intend concrete structure henceforth.[2]

What about the term "relata"? This is especially important, in part because there is a significant degree of ambiguity regarding this term in the literature on SR. The relata that most philosophers have in mind in this context are putative objects: fermions; molecules; human beings; etc. And one can offer canonical examples of these things for illustrative purposes: the electron; the hemoglobin molecule; Socrates. Though there is nothing wrong with the common practice of taking the putative relata in these debates to be objects, it is also potentially misleading. Commonly, when speaking of objects, we have the referents of count nouns in mind—things that can be counted, or individuals. In the context of structuralism, however, this common conception of objects is too narrow, because the putative relata of scientific relations are often not associated with count nouns at all, but mass nouns—things that cannot be counted but merely quantified, like plasma, or kinetic energy.

If OSR is to be a compelling view, it must apply to both individuals and non-individuals, count nouns and mass nouns alike. Thus, for the purposes of this discussion, I will leave aside the issue of whether or not there are individual objects, which confuses much of the literature; the relevant contrast here is between structures and non-structure, or relations and relata, not between structures or relations on the one hand, and individuals on the other. I will construe the term "object" broadly here so as to include all sorts. An object is anything associated with a group of properties that cohere at a location. As a final clarification, let us note that the relata of scientific relations are often not objects at all, but rather properties of objects, as in the example of the ideal gas law.

---

[2] For a more detailed discussion of the notion of concrete structure, see [6, e.g. pp. 40–41].

10.2 Theories of Object Ontology

With these clarifications of the concepts of relation and relata in hand, let us now return to the idea of OSR in more detail. Recent work, particularly in the philosophy of physics, has fuelled an impressive proliferation of positions claiming to be versions of OSR. My present aim is to construct a general argument regarding this class of views as a whole; thus it will serve us to begin with a rather general characterization of it in order to collect all of the members of this class together. Taking into account the various options now proposed, Ladyman [18] describes OSR generically as follows: "On the broadest construal OSR is any form of structural realism based on an ontological or metaphysical thesis that inflates the ontological priority of structure and relations." Objects and properties are often and traditionally described as having a kind of basic or primary ontological status, whereas relations and the structures they compose have a derivative status. OSR broadly construed, however, seeks to reverse this thinking: it treats structural relations as primary, and objects and properties as derivative (at best).

This raises many questions, but to begin, note how much broader this formulation is in comparison to the original description of OSR in eliminative terms. According to eliminative OSR, there are structures in the world, but nothing that could be correctly described as objects or properties standing in structural relations. This take on OSR is certainly consistent with the newer, broader formulation, but it would appear that it is not unique in this respect, prima facie. "Inflating the ontological priority of relations" here means inflating the priority of relations relative to that of their putative relata, and clearly, denying the existence of the relata altogether is one way to achieve this. There would seem to be other, less strong medicines, however, with which to produce the same result, for at first blush, the idea of greater "priority" is *also* consistent with the idea that one thing is ontologically more basic or fundamental than something else whose existence is not in question. The standard metaphysical test for determining how fundamental something is, relative to something else, is to think about relations of dependence that may exist between them. Individual birds are ontologically more basic than populations of birds, for example, because populations depend for their existence on the organisms that compose them, and not vice versa. I will have more to say about the crucial idea of ontological priority momentarily. For the time being, however, let us proceed with the idea that, on its newer and broader construal, OSR would seem to include any form of SR according to which the relata are in some sense ontologically dependent on relations involving them.

In order to consider the full range of conceptual possibilities for OSR thus broadly construed, let us start by imagining an ontological spectrum of conceptions of the relata, ranging from "thick" conceptions at one end to "thin" conceptions at the other. Figure 10.1 represents a comprehensive mapping of theories of object ontology along this spectrum. At the thick end we have metaphysical theories that give high ontological priority to objects, and relatively less priority to the relations in which they may stand. In the limit at the thick end we have realism about substance: a metaphysical commitment to brute, primitive principles of objecthood. Typically

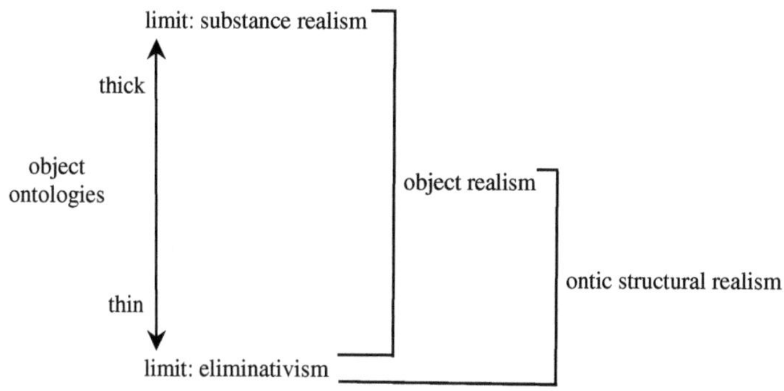

Fig. 10.1 Object ontologies and the ostensible boundaries of OSR broadly construed

on such views, objects are composed of bare substrata, the very concept of which defies further analysis. Properties inhere in or are instantiated by substrata, forming composites, and relations obtain between these composite entities, inter alia. At the other end of the spectrum is eliminativism, the view that there are no such things as objects (or properties) at all.

As noted above, OSR was originally identified narrowly with eliminativism, according to which the notion of an object is simply a kind of metaphysical illusion, to be jettisoned once we have a better understanding of the fundamental nature of reality as revealed by physics. It is for this reason that in these earlier days, Stathis Psillos [23, pp. S18–S19] labelled the position "eliminative structural realism", in contrast to the label "restrictive structural realism", which he gave to ESR. Still true to the spirit of pure eliminativism is Steven French [15], for whom talking about objects is simply a *façon de parler*, a useful manner of speaking forced on us by the subject-predicate form of our natural and standard logical languages. This form is misleading, he maintains, and the putative relata of structural relations are merely artefacts of this manner of speaking. Today, however, on the broader construal, OSR has diversified; the position now includes but is not restricted to eliminativism. A number of authors including Simon Saunders [26, p. 163], Tian Yu Cao [5, p. 41], John Stachel [29, pp. 52–58], and Ladyman [19] have recently suggested that things such as space-time points, quantum mechanical particles, and other putative objects really are objects, but objects whose natures depend on and are determined by the relations in which they appear. It is the relations that have ontological priority.[3]

In the following two sections, I will examine the spectrum of theories of objects mapped in Figure 10.1, from thick through increasingly thin conceptions, with the aim of determining which if any hold promise for a defensible account of non-

[3] Ladyman and Ross [20] are clear that their position is a form of eliminative OSR, endorsing the thesis that "there are only relations, and no relata" (e.g. pp. 151–152). As we shall see, however, they sometimes appear to endorse non-eliminative OSR. Some authors advocate forms of non-eliminative OSR on which neither relations nor relata have ontological priority. I will consider this possibility in Section 10.5.

eliminative OSR. This promise, as we shall see, turns on the issue of whether an appropriate understanding of ontological priority can be made to fit the bill. Let us turn to this issue now.

10.3 Establishing the Ontological Priority of Relations

The idea of ontological priority is at the very heart of distinctions between various forms of OSR, and also a point of contention between OSR and more traditional views of both scientific realism and the metaphysics of objects. Ontological priority is supposed to concern how basic or fundamental something is relative to something else, but what does this mean, more precisely? Recall the suggestion earlier that the standard test for determining how fundamental things are with respect to one another is to cite relations of dependence between them. This is an unhelpfully vague suggestion as it stands, however, because it is not entirely clear in this context what the relevant relation of dependence *is*. Indeed, we commonly speak of different kinds of metaphysical dependence—most commonly mereological, modal, causal, and supervenient—as being variably exemplified in different contexts. Thus, in order to understand more clearly here what it means for one thing to have ontological priority over something else, we had better be explicit about the specific relation of dependence that is relevant to the context of non-eliminative OSR. The clarification required thus takes the form of (1) below. Taking *R* to be an appropriate relation,

(1) *x* is less fundamental than y with respect to *R* iff *x* depends on *y* with respect to *R* and not vice versa.

The clause "and not vice versa" appears in the second part of the biconditional in order to rule out cases in which *y* also depends on *x* with respect to *R*, in which case it would seem correct to say that neither *x* nor *y* has ontological priority relative to the other. In the example I gave earlier, the fact that populations are less fundamental than organisms—with respect to the relation of composition—can be inferred from the fact that populations depend on organisms with respect to composition and not vice versa. Populations are composed of organisms and not the other way around.

The present investigation concerns the relative ontological priority of scientific relations and relata. What is the relevant relation (or relations) of dependence according to which priority should be assessed in this context? This, I submit, is not a question with an obvious answer. On the eliminativist approach to OSR, there is of course no need to answer this question, because on this approach there are simply no objects to be considered more or less ontologically fundamental than relations. By eliminating objects from their ontology, proponents of eliminative OSR are spared the task of specifying *R* in (1), but as noted previously, at the cost of having to meet another serious challenge: that of making intelligible the idea of concrete relations in the absence of relata. The non-eliminative approach to OSR, conversely, ostensibly escapes the challenge of having to make sense of concrete relations without relata, because it admits the relata into its ontology, but then must face up to the question just posed: in what sense are relations more fundamental? Scientific theories quantify over a staggering range of objects and properties, and clearly not *all* of the relations between these entities are mereological, or modal, or causal, or relations of supervenience. Is there any relation of dependence that is sufficiently general to serve as *R* in (1) for non-eliminative OSR, or is its thesis of ontological priority inevitably highly disjunctive?

As it happens, there is at least one relation that would appear to be adequately general here. If there is a common thread running through the literature on non-eliminative OSR, it is the idea that the *identities* of scientific relata, whether objects or properties, are in some sense determined by the relations that obtain between them. That is to say that the natures of the relata—whatever it is that makes them what they are—are not properly understood in terms of their intrinsic features, if indeed they have any. Rather, their natures are a function of purely extrinsic features: their relations. As Saunders [26, p. 163] describes it, it is often reasonable to say that "a particular body is no more than a particular pattern-position"; or in the words of Ladyman and Don Ross [20, p. 131], "there are objects in our metaphysics but they have been purged of their intrinsic natures, identity, and individuality, and they are not metaphysically fundamental". Despite their differences, the various proponents of non-eliminative OSR thus appear to share the view that "all there is" to certain objects are the relations in which they stand. The relation of dependence relevant to the context of non-eliminative OSR is thus the relation of (what I will call, somewhat awkwardly) the determination of identity: certain relations described by scientific theories determine the identities of at least some of the objects or properties that stand in those relations. To put this into the form sketched in (1) for understanding the notion of ontological priority, let me render the idea as follows:

(2) The relata are less fundamental than their relations with respect to the determination of identity iff the relata depend on their relations for the determination of their identity and not vice versa.

With this explication of ontological priority in hand, let us now consider more precisely the sense or senses in which the identity of an object or property might be determined by its relations. To begin, recall the space of conceptual possibilities mapped in Fig. 10.1. Starting at the thick end of the spectrum, it is immediately obvious why theories of object ontology positing substances are incompatible with OSR. Historically, the concept of the bare substratum was introduced inter alia for the express purpose of accounting for the identities of objects. Properties and relations may come and go, but the anchor of identity, on this view, is a primitive feature of the bare substratum: its haecceity or primitive thisness. Having no qualitative natures, haecceities are truly mysterious; they are purely and simply principles of identity. The thoroughly opaque nature of these principles, being as they are impervious to scientific or other empirical investigation by definition, has long been cited as a reason for dismissing the plausibility of theories invoking them. The important point for present purposes, however, is simply that any such theory is incompatible with OSR, because on such a view, objects have too high a degree of ontological priority. On such a view, concrete relata cannot be less fundamental with respect to identity than the relations in which they may stand, because their identities are completely independent of such relations.

Let us thus move along the spectrum presented in Fig. 10.1, away from theories of substance, in the direction of metaphysically thinner views of the nature of objects. Object ontologies that repudiate substances are generally versions of the bundle theory: the idea that objects are simply groups of properties that cohere at locations in space-time. From the perspective of non-eliminative OSR this is certainly an improvement, but it does not yet go far enough, for although on such a view we have now set aside the notion of substrata and their intrinsic principles of identity, we have not yet gone so far as to refrain from emphasizing intrinsic natures. As bundle theories are usually interpreted, the properties of an object that are taken to determine its identity, though not inhering in a substratum, are nonetheless intrinsic to the bundle constituting it. Thus, here again we find the relative ontological priority of relations and relata skewed in favour of the relata, for again, the identities of objects are determined by something intrinsic, as opposed their extrinsic features, *viz.* their relations. What is required for the purposes of non-eliminative OSR is an account of the relata that emphasizes their relational features, and this requires a move to an even thinner conception of objects. In the next section, I consider one such possibility.

10.4 Thinning Out with the Dispositional Identity Thesis

The task before us is to come up with a coherent metaphysical picture of objects whose adoption would shift some significant ontological weight—as much as possible, for the sake of non-eliminative OSR—from objects to the relations in which they stand. The only well-developed option among extant views that places significant emphasis on relations in connection with questions of identity is what I will refer to as the dispositional essentialist view of properties; it is sometimes called a "structural" view.[4] The basic idea is as follows. Consider the nature of physical or causally efficacious properties of concrete objects, as opposed to logical or mathematical properties. To say that an object has a property of the former sort is to say that it is disposed to behave in certain ways in certain circumstances. That is, in the presence or absence of other properties and objects, it will stand in certain *relations*. A property on this view is identified as the property that it is in virtue of its possible relations to other properties. The conjunction of all possible relations thus comprehensively describes the natures of all properties.

A clear statement of the general idea of dispositional essentialism is given by Sydney Shoemaker [27, p. 133], who claims that "the identity of a property is completely determined by its potential for contributing to the causal powers of the

[4] This account of the nature of properties can be traced to Shoemaker [27] and Swoyer [30], and has been discussed in significant detail more recently by a number of authors, including (for example) Hawthorne [16], Bird [3], and Chakravartty [6, chapter 5].

things that have it". The term "causal power" here is for all intents and purposes synonymous with what I intend by the term "disposition" (the differences are minor and in any case, immaterial presently). I call this view dispositional essentialism because it maintains that what makes a property the property that it is, or in other words, what constitutes the essence of a property, are the dispositions for relations it contributes to the objects that have it. It is now obvious, perhaps, why such a view might be tantalizing for a structural realist concerning scientific knowledge, and more specifically, for an advocate of non-eliminative OSR. If we were to marry this structural view of properties to the view that objects are simply groups of properties that cohere (that is, if we combine it with a rejection of substances), we would then have the makings of an account of properties and objects that emphasizes relations very significantly in giving an account of identity. On this combined view, the very natures of properties are understood simply in terms of potential relations, and objects are simply groups of properties.

The main alternative to the dispositional essentialist view of property identity is what David Armstrong [2, pp. 26–27] calls a "categoricalist" theory of properties, according to which they have "a nature of their own" quite independently of anything having to do with their relations. The idea of a primitive principle of property identity, or quiddity, is analogous to the idea of haecceity or primitive this-ness in connection with the notion of bare substrata. Just as in the case of haecceities, many have argued against quiddities on the grounds that they are fundamentally mysterious and entirely impervious to scientific or other empirical investigation. To be fair, Armstrong [1, pp. 168–169] explicitly denies that quiddities are required to account for property identity on his view. Instead, he maintains, one may simply stipulate that different properties are numerically distinct. It is arguably unclear, however, how brute numerical difference is any less mysterious than difference by quiddity (Armstrong suggests that one might think of the former as "a difference in another 'dimension', orthogonal to the dimensions of spacetime"). In any case, given the current task of scouting potentially compatible accounts of objects for non-eliminative OSR, and given that in this connection we have already rejected substances and primitive identities, it seems we have no choice but to reject categoricalism about properties too, for much the same reason: it adopts a principle of identity that is incompatible with the idea that relations have greater ontological priority than their relata.

The turn to a structural or dispositional essentialist understanding of the natures of properties emphasizes the relations in which they stand, and it is for this reason that one might reasonably think it suitable for a non-eliminative, ontic structuralist conception of objects. Now the bad news: the compatibility of this view of properties and non-eliminative OSR, it turns out, is only skin deep. For while it is true that dispositional essentialism emphasizes relations as opposed to primitive and mysterious intrinsic features in giving an account of property identity, it does not emphasize relations in quite the right way for OSR. On the dispositional essentialist view, it is not relations per se that determine the identities of properties, but rather the generally intrinsic *potential for* relations. That is what a disposition is: it is a causal power, and the causal powers investigated by the sciences are generally intrinsic properties.[5] Of course, we often *describe* dispositions in terms of their manifestations—that is, in terms of relations—but it would be a mistake to conclude on this basis that such relations constitute their conditions of identity. The intrinsic dispositions of objects exist quite independently of whether or not they are manifesting—in other words, independently of whether they or the objects that have them are standing in any particular relations at any given time. Therefore, on this view, it is simply incorrect to say that the relata depend on their relations for the determination of their identity, and thus, recalling (2), we are not entitled to infer that the relata are less fundamental than their relations.

So close, and yet not close enough. The lure of dispositional essentialism is certainly seductive for the non-eliminative approach to OSR. Indeed, this view of properties places so much emphasis on relations that it engenders a kind of holism regarding the natures of properties. If the identity of a property is determined by certain dispositions for relations with other properties, it would seem that the natures of properties taken as a whole are constituted by a vast network of potential relations. The natures of individual properties are thus linked to one another via loops of potential relations. Despite the appeal to relations in giving this account of property identity, however, it remains the case that on this view, identity is determined by the potential for such relations, not the relations themselves. The potential for relations is encapsulated in the concept of a disposition, which here applies to intrinsic properties of objects. On this view it is simply not the case that the identities of objects depend on the relational structures of which they are part, or as Michael Esfeld [9, 10] puts it, that object identity is determined purely by means of relational properties. The dispositional essentialist or structural view of properties and objects is not thin enough for non-eliminative OSR. We must get even thinner.

10.5 Exclusive Disjunction: Eliminativism or Intrinsic Identity

The goal from the outset has been to identify an account of object ontology that would serve to make sense of OSR without lapsing into eliminativism, but the conceptual space in which to locate such a view has been shrinking with each successive consideration. It would seem that the only course remaining is to explore the space left, in Fig. 10.1, between two possibilities. The first is the possibility we have just considered, *viz.* the combination of a bundle theory of objects and a view of properties that describes their identity conditions in terms of dispositions for relations. The second possibility is the limiting case of thin accounts of objects itself: eliminativism. In this penultimate section I will argue that having come this far, we have simply run out of room—what little conceptual space there remains to explore in Fig. 10.1 provides no refuge for non-eliminative OSR.

[5] There are also such things as extrinsic dispositions, but their existence is inconsequential to the point here. I consider this issue in Section 10.6.

To recall the desiderata, we are attempting to identify a theory of properties and objects that is compatible with the view that these entities are ontologically less fundamental than their relations. A theory meeting this description should entail that the relata depend on their relations for the determination of their identity. In light of considerations outlined in the previous section, we also know that the theory we seek must link the identity conditions of properties and objects to the relations in which they stand even more directly than the link described by the dispositional essentialist view of properties. Throughout this discussion a particular difficulty has persisted: an inability to liberate objects from their intrinsic properties. So long as this difficulty persists, the prospects for non-eliminative OSR seem grim, for so long as the relata of the relations described by scientific theories have genuinely intrinsic features, it seems impossible to satisfy the second part of the biconditional in (2). Recall that this part asserts that the relata depend on their relations for the determination of their identity and not vice versa. So long as the relata *have* genuinely intrinsic features—qualitative properties, dispositions, what have you—this condition remains unsatisfied, because these intrinsic features keep popping up as plausible candidates for determining their identity. On the various accounts of object ontology we have surveyed, there is no respite from the intrinsic.

Thus, let us slim down even further and consider the possibility of an ontological theory of the relata according to which they have no intrinsic features at all: no qualitative intrinsic properties; no intrinsic dispositions; nothing intrinsic that would admit of any sort of description on the basis of scientific or other empirical investigation; and no haecceities or quiddities either. This is to imagine the possibility of a theory permitting nothing in terms of which object identity could be determined intrinsically. But now, I believe, it should be clear that we have come too far, because an object with no intrinsic features at all, whether knowable or unknowable in principle, is not an object at all. Lacking *anything* intrinsic—no substratum, no properties, no primitive principles, or what have you—there is simply nothing left to stand in any sort of relation. In the attempt to locate a metaphysical theory of objects that is compatible with non-eliminative OSR, we have crashed with a bang into the limit of eliminativism, and there does not appear to be any way to apply the brakes sooner. Concrete objects that have no intrinsic features are not anything, and once we have gone this route, we have embraced eliminativism. I suspect that this contention will seem intuitively obvious to some, and generate worries about begging the question in others. Let me elaborate the contention below.

The idea that object identity is purely extrinsic has a natural home in the philosophy of mathematics, and the hope that this idea can be pressed into service in connection with concrete objects appears to be a consequence, in at least some cases, of a conflation of the objects of mathematical and scientific investigation. Randall Dipert [7], for example, argues for the purely extrinsic identity of objects on the basis of his conviction that the world *itself* is a mathematical object. Echoing the arguments of structuralists about mathematical entities, he contends that if one accepts that the world is a mathematical structure, the identities of the objects and properties populating it can be analyzed perspicuously using graph theory. Graphs are mathematical structures composed of two things: nodes (or vertices); and edges between nodes, which can be taken to represent relations. These relations may be directed, in which case the relation between two nodes joined by an edge is taken to be asymmetric, or undirected, in which case such relations are symmetric. Graphs may be labelled or unlabelled; in the former case linguistic or numerical labels are assigned to the nodes. The interesting point for present purposes is that some graphs are themselves asymmetric: they are arranged in such a way that each of their nodes is related to the others in a unique manner, thus facilitating wholly extrinsic assignments of identity. "In an asymmetric graph, it is possible to give a unique, purely structural description for each vertex" [7, p. 348]. In Dipert's view, the world is an asymmetric graph.

Is this at all convincing, or is it procrustean? Structuralism is a subject of debate even where it is most plausible, regarding mathematical entities in the philosophy of mathematics, and its adoption in that context is by no means universal, but let us leave this point aside here. Granting for the sake of argument that purely extrinsic identity is an ultimately defensible view in the philosophy of mathematics, are there grounds for thinking that the world is an asymmetric graph? No such grounds have emerged. What would be required, in the first instance, is a compelling argument to the effect that reality as we know it is a mathematical object, and in the second instance, one might reasonably require a compelling argument to the effect that this reality is correctly described as an asymmetric graph. But arguments for the thorough-going Platonism or Pythagoreanism exemplified by the first proposition are subject to long-standing objections, generally regarded as fatal, and non-eliminative OSR issues no new arguments on its behalf. As an *assumption* about the nature of the world of the concrete offered *ex cathedra*, the view is thus rather lacking in motivation. Though we rightly entertain lively debates about what it is to be (for example) a subatomic particle or an organism, in the sense of having a defensible ontological theory, any view so amazing as to entail that particles and organisms (for example) are purely mathematical entities, in the absence of strong motivation, presents itself as a *reductio*.[6]

Ladyman [19] also appeals to graph theory, as an analogy, in hopes of motivating the idea of objects whose identities depend solely on the relations in which they stand. Figure 10.2 depicts a labelled, asymmetric graph whose properties he considers (p. 36). Each of its nodes can be uniquely identified purely on the basis of its relations to other nodes, and they are labelled so as to illustrate this in the following way: each node is described by a list of numbers, one number for each node to which it is directly related; the particular number assigned is given by the number of nodes to which the latter is or are related. Consider the node uppermost in Fig. 10.2, for example. It is directly related to three other nodes, hence its label consists of three numbers. The node beneath it towards the left is related to two others, the node beneath towards the right is related to three others, and the node

[6] Aware of this difficulty, Ladyman and Ross [20, p. 158] thus reject the mathematical characterization of concrete reality. But their rejection goes only so far: "What makes the structure physical and not mathematical? That is a question that we refuse to answer."

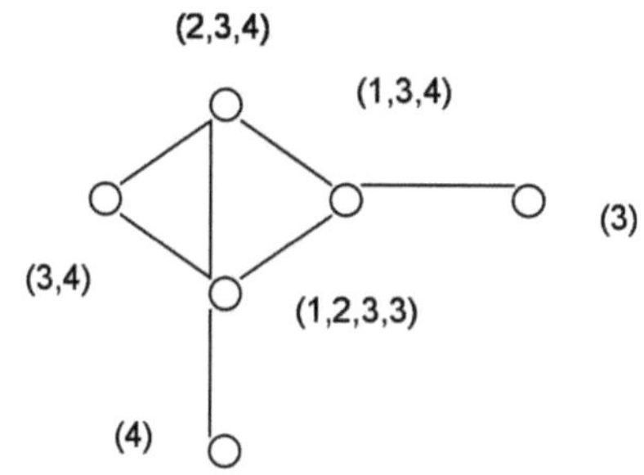

Fig. 10.2 An example of extrinsic identity from graph theory

directly underneath is related to four others, generating the label "(2,3,4)", which is unique to the uppermost node.[7]

Now let us pose the question the analogy suggests: does this mathematical object constitute a model that might help to clarify the idea of purely extrinsic identity in the natural world, the world of concrete objects and property instances described by the sciences? In the absence of something like the thorough-going Platonism or Pythagoreanism of Dipert, the answer, I submit, is no. Let us imagine that we were able to draw the diagram of a graph purporting to represent the relations between a number of objects or property instances constituting a target system of interest in the natural world. Let us also assume that each of the nodes of this graph is uniquely identifiable purely on the basis of the structural features of the graph as a whole. Even with these riches in hand, we would not then have a principle of extrinsic identity on the basis of which to make sense of non-eliminative OSR, and the illustration in Fig. 10.2 is helpful in facilitating a demonstration of why this is so.

Absent Platonist or Pythagorean extremism, there is a crucial difference between a graph *qua* mathematical entity and a graph *qua* representation of some aspect of the natural world, and this difference fatally undermines the analogy of graphs to concrete systems with respect to identity. Unlike the case of graphs considered abstractly in purely mathematical terms, when graphs are employed to represent the concrete, their nodes are occupied, ex hypothesi, by objects and/or property instances. That is to say, they denote other things, external to themselves. But the labels in graphs such as the one illustrated in Fig. 10.2 do not furnish identity criteria for objects or properties that might *occupy* their nodes in such a representation; they simply identify *locations* in a structure! There is a crucial difference between identifying a location in a structure and constituting the identity of something that occupies that location. Graph labels may furnish extrinsic identity for locations in mathematical structures, but this does not by itself suggest anything at all about the identities of the sorts of things that stand in most structural relationships of

[7] The analogy here suggests viewing objects or properties as individual nodes. Alternatively, Dipert [7] holds that an object is "a subset of the vertices of the world graph" (p. 352), and that even fine-grained entities such as subatomic particles are not vertices but "composite entities, subgraphs of the world graph" (p. 356).

interest to the sciences. Unlike nodes, such objects are generally described as having intrinsic properties, which present themselves as plausible aspects of identity. Graph theory would only provide a helpful analogy here if we were to think of the world as exhaustively comprised of purely mathematical entities such as nodes. But alas, absent Platonist or Pythagorean extremism, concrete objects and mathematical objects are different in kind, and the sciences are interested in the former and not merely the latter.

Where does this leave us? The recent evolution of OSR as a proposal for scientific realism has led inexorably to an interesting but conceptually fraught disjunction. On the original, eliminative formulation of OSR, the ontology of objects and properties is of little concern, since on this view, there are no such things. Indeed, nothing I have said in this essay tells against eliminative OSR, but the challenge facing this position, to make intelligible the idea of concrete relations in the absence of ontologically significant relata, remains. The revisionist metaphysic according to which such relations may be viewed as ontologically subsistent in their own right, many suggest, has not yet arrived.[8] On the other hand, the non-eliminative position evades the challenge of explaining how there can be concrete relations without relata, but consequently faces another challenge—that of making intelligible the idea that such relata have less ontological priority than the relations in which they stand. I have argued that the analogy to the notion of extrinsic identity in the context of mathematical entities offers no help in this regard. The upshot of these considerations is, I believe, an important conclusion about the prospect of a tenable OSR: apparently one cannot deny eliminativism *and* a role for the intrinsic in the analysis of identity in connection with the relata of scientific discourse; if we deny one, we are stuck with the other.

Is there any way of escaping this dilemma by weakening OSR further? Some recent proponents of OSR have suggested yet another interpretation of the view, according to which structuralists should both grant the existence of objects and *relax* the condition that they have less ontological priority than the relations in which they stand. This relaxation cannot extend so far as to give greater ontological priority to objects, of course, on pain of dispensing with OSR altogether, but why not, so the suggestion goes, afford relations and relata the *same* ontological status, granting neither ontological priority with respect to the other?[9] Given that traditionally, relata are usually understood to have greater priority, it would seem that this prescription also satisfies, if only just, the broadest construal of OSR with which we began, which embraces any form of SR based on an ontological thesis that "inflates" the ontological priority of relations with respect to their relata. It should be clear immediately,

[8] For critical discussions of eliminative OSR, see [4, 24], which express different concerns regarding Platonism in that context inter alia, and [6, pp. 70–85]. For more recent developments of the view, see [15].

[9] French [14] traces this idea as a proposal for OSR to Eddington's view of subatomic particles (the only alternative Eddington considers is an object ontology involving substances). A similar view is described by Pooley [22, p. 98], and endorsed by Esfeld [9, 10], Rickles [25, pp. 188–191], Esfeld & Lam [11], and Floridi [12, pp. 235–236].

however, that this suggestion can fare no better than the version of non-eliminative OSR we have already considered. For insofar as it attributes nothing intrinsic to the relata of scientific relations, it slips without acknowledgment into eliminativism: lacking anything intrinsic, there remains nothing between which relations could obtain. If instead one attributes intrinsic properties (or other intrinsic features) to the relata, the intrinsic then constitutes a plausible determinant of their identity. Lacking an account of extrinsic identity for concrete objects, there is simply no metaphysical space, it seems, capable of sheltering a non-eliminative OSR.

Earlier I suggested that stripping objects of every vestige of intrinsicality (primitive attributes, qualitative properties, and so on) is tantamount to eliminativism, and that while this contention will seem intuitive to some, it may appear question begging to others. For why not simply accept, the latter might contend, that an object can have only extrinsic properties? This, I submit, will not do. There is an important conceptual asymmetry between the intrinsic and the extrinsic in this regard. As they are usually parsed, intrinsic features are, in the jargon, ones that are possessed independently of accompaniment or loneliness. To put it figuratively, they are "contained within" an object, and thus (in part or wholly) constitute it. In contrast, an extrinsic feature is one that is possessed by an object in virtue of its relation (or relations) to some other thing (or things), and therein lies the rub. What is the "it" in the phrase "its relation to some other thing"? If the answer is to be given purely extrinsically, one is left with a circularity or regress. For then, in order to comprehend what the "it" is in the context of this purely extrinsic characterization of the object, one has no option but to appeal to the extrinsic once more. And so on. In order to break this cycle and thereby give content to the notion that there is something that has the relevant extrinsic properties, one must first grasp the idea that there is a something that may enter into a relation, before then proceeding to entertain the idea that it does. The very attribution of an extrinsic property assumes that one has a prior grasp, ontologically speaking, of what it is that stands in the relevant relation or relations.

There is no parallel difficulty in the context of intrinsic features, since here the "it" is simply the collection of these features, which can be understood to stand in relations to external things with no threat of circularity or regress. Extrinsic properties thus cannot by themselves constitute objects in the way intrinsic properties do, and the notion of an object consisting solely of extrinsic properties is, at the very least, a serious conceptual puzzle. Perhaps this puzzle can be solved, but in the meantime, it would seem that if non-eliminative OSR is to adopt the view that objects are purely extrinsic in nature, the conundrum it faces is no less profound than that facing eliminative OSR. Indeed, as I have suggested, it amounts to the same thing. On this interpretation the position is thus a version of eliminativism, sharing whatever promise or difficulties the latter view engenders.

10.6 Extrinsic Dispositions and Lessons from Physics

Thus far, the argument of this paper has been conceptual. The literature advocating non-eliminative OSR is rife, however, with claims to the effect that plausible interpretations of our current best theories in physics *demand* that one accept the ontological priority of non-eliminative structural relations over their relata, or the no-priority thesis—precisely the views I have dismissed on conceptual grounds. A detailed refutation of each of these more specific claims would, in fairness to them, require a series of essays engaging the case studies in which they are immersed. Nevertheless, in this final section, I will offer some principled reasons for thinking that all of these more specific claims are, quite generally, susceptible to the arguments I have outlined above.

First, let me set one important issue, which inevitably arises in this context, to one side. Arguments for the view that plausible interpretations of modern physics point towards non-eliminative OSR focus specifically on interesting facts about quantum theory and general relativity. It is commonplace among philosophers of other sciences to wonder why these studies should be thought to yield *general* morals regarding ontological priority, not least because objects and properties in other domains of scientific theorizing, though subject to interesting philosophical puzzles of their own (including ones concerning identity), do not seem to require or even suggest any deep revisions to our views regarding the relative ontological priority of objects and properties on the one hand, and relations on the other. Suggestions to the effect that basic physics provides general morals of this sort must, it seems, appeal to forms of reductionism that many find implausible ("there are no objects or properties other than those described by basic physics"), or arguably beg the question [20, p. 44] ("primacy of physics constraint": "Special science hypotheses that conflict with fundamental physics... should be rejected for that reason alone."). I am sympathetic to these concerns, and suspect that neither reductionism nor assuming the primacy of physics amounts to a compelling basis for arguments about ontological priority across the board, but let us leave these issues to one side for present purposes.

My reasons for not entering into debates about the ultimate scope of lessons from basic physics here are twofold. For one thing, though they are interesting in their own right, these debates are couched in discussions of rather different issues than those I have considered here, including disputes about reductionism and the unity of science. Secondly and more importantly, entering into these debates in the present context is supererogatory, metaphysically speaking. For even the restricted domains of physics within which arguments for non-eliminative OSR arise are subject to the philosophical considerations I have marshalled above. Since claims about how physical descriptions of the nature of quantum particles and space-time points lead irresistibly to non-eliminative OSR are *themselves* mistaken, there is no question of extending the moral of non-eliminative OSR to other regions of the sciences. I will comment briefly on these claims in turn, considering first the argument from quantum theory, and then the argument from space-time physics.

To begin, let me recall part of the discussion in Section 10.4 concerning the dispositional essentialist or structural view of properties. Given the emphasis this view places on the potential relations of objects in providing an account of property identity, it seemed initially to offer some support for the idea that objects depend for their identity on the relations in which they stand. This promise was short-lived, however, in light of the observation that property identity on this view is not conceived in terms of relations per se, but rather in terms of dispositions for relations, which are generally described as intrinsic properties where scientific entities are concerned. There are, however, such things as *extrinsic* dispositions, and this might be thought to open the door a crack to the possibility of non-eliminative OSR as an appropriate view of at least some scientific objects. If it could be shown that there are objects whose properties are described exclusively in terms of extrinsic dispositions by the relevant branch of physics, one might then at least have a pressing motivation to overcome the conceptual puzzle described at the end of Section 10.5, regarding the idea of purely extrinsic identity.

An extrinsic disposition is an extrinsic property: one whose possession by an object depends on something (or things) external to itself; it is possessed in virtue of some relation or relations to that thing (things). Of course, establishing that a disposition is in fact extrinsic is not always straightforward. Since we routinely describe intrinsic dispositions in terms of possible manifestations in certain circumstances—circumstances that are usually extrinsic to the object in question—a great deal of description of that which is external to an object may be applied in ascribing even its intrinsic dispositions. Consider the intrinsic disposition of solubility. This is an intrinsic property of solutes, but it is usually described in terms of its characteristic manifestation, dissolving, which occurs in some external circumstances (being placed in an appropriate solvent having an appropriate degree of prior saturation at an appropriate temperature, and so on) and not in others. Many of what might at first glance appear extrinsic dispositions are in fact intrinsic; their possession by an object is independent of its external circumstances, though their manifestations are not.[10]

Quantum theory provides an example of what may be interpreted as genuinely extrinsic dispositions, and this is what fuels the claim that a plausible interpretation of the theory supports a version of non-eliminative OSR. Particles (two electrons, say) described by the theory as entangled bear relations to one another that, on an orthodox interpretation of the theory, do not supervene on any intrinsic properties of the particles themselves. These relations of entanglement are interestingly correlated with measurement outcomes for the values of certain properties such as particle position, momentum, and spin. Quantum theory does not ascribe determinate values of these properties to the particles, but rather describes only correlations between them by means of a joint probability distribution determined by their joint state. Thus, here it seems we have a prima facie case for a disposition of an object—to be measured as having certain values for certain properties—that is wholly independent

[10] See [21] for a defence of the idea of extrinsic dispositions. Though some of the examples presented are arguably intrinsic dispositions described in terms of extrinsic manifestation conditions, others are clearly extrinsic.

of its intrinsic properties: an extrinsic disposition. There are other interpretations of quantum theory that describe relations of entanglement as supervening on intrinsic properties, each of which comes with the price tag of an arguably unpromising metaphysical supposition (the existence of superluminal interactions, backwards causation, and other possibilities), but let us grant for the sake of argument here the orthodox interpretation according to which quantum particles have relations that cannot be analyzed in terms of anything intrinsic. Would this demonstrate that the identities of these objects are extrinsic, in the manner suggested by the ontological priority thesis of non-eliminative OSR?

The answer, I suggest, is no. In order for the appeal to extrinsic dispositions in this context to offer any support to non-eliminative OSR, it would seem that one of the two following conditions should obtain. Either it should be the case that not just some, but *all* properties of the particles described by quantum theory are extrinsic, or it should be the case that whatever intrinsic properties the theory does attribute to them do not determine their identity. If either of these conditions were to hold, purely extrinsic identity might seem a natural hypothesis to explore, and the motivation to overcome the conceptual difficulties presented by non-eliminative OSR would intensify. But neither condition obtains. While some properties are described by quantum theory in terms of relations of entanglement, others are not. Mass and charge, for example, are state-independent intrinsic properties of subatomic particles, whose attribution thus violates the first condition that all properties be extrinsic. Furthermore, if such entities are to be the sorts of entities they are, they must instantiate particular values of these properties, thus violating the second condition that intrinsic properties do not determine identity. The identity of quantum mechanical entities is thus not extrinsic. Much is made in debates surrounding OSR of the peculiarity that quantum particles cannot be individuated on the basis of their intrinsic properties, but to cite this as evidence in favour of non-eliminative OSR is to conflate individuation with identity, and this is at best controversial. The individuation of quantum particles is a thorny issue, but whatever one makes of it, the fact remains that wherever one applies the concept of a particle, the theory presents descriptions of what appear to be intrinsic properties which are constitutive of their identity.

Similar considerations apply to the case of space-time physics, the other arena in which some have argued in favour of augmenting the ontological priority of relations with respect to their relata.[11] One might wonder, for example, whether space-time points are objects, and if so, whether they have identities that can be understood purely extrinsically, in terms of spatio-temporal relations. Esfeld and Vincent Lam [11] answer yes to both questions and offer this as a version of non-eliminative OSR, according to which "an object as such is nothing but what bears the relations" (p. 31). But again the question arises: what is the "what"?—that is to say, what is the thing that bears the relations? If there is nothing intrinsic, there would appear to be

[11] There is no consensus among structuralists regarding how to apply SR to general relativity. Distinct from the approach of non-eliminative OSR considered here, for example, Dorato [8] views SR as furnishing a third option in debates about relationism and substantivalism, and Slowik [28] sees it as a means to avoiding these debates altogether.

nothing relative to which anything could be extrinsic. An answer to this conundrum is owed, but none is presently forthcoming on behalf of non-eliminative OSR. On the assumption that space-time points are objects, there is a fact of the matter about whether these objects have intrinsic properties or other intrinsic features. If they do, then it is reasonable to suppose, as in the case of quantum particles, that their identities are not purely extrinsic, quite independently of the question of whether their individuation requires recourse to various relations.

Lest the arguments of this paper be interpreted too strongly, let me close by noting that there may well be instances of extrinsic identity in special cases. For example, according to the phylogenetic species concept, what makes me a member of the species *Homo sapiens* as opposed to the species *Homo neanderthalensis* is a particular relation of descent I bear to a particular hominid ancestor; this is constitutive of my identity as a human being. This makes the prior assumption, however, that there is such an object as me—recall that the attribution of an extrinsic property, such as being descended from an early hominid, assumes some prior ontological grasp of that which stands in the relevant relation, on pain of circularity or regress—and this presupposition is generally explicated in terms of my intrinsic properties, differences in which help to distinguish me from my colleagues down the hall. Similarly, what makes something a space-time point (if there are such things) as opposed to a subatomic particle is a function of some important intrinsic differences, even if it turns out that in order to individuate one space-time point as distinct from another, one must rely on their extrinsic properties. The existence and role of the intrinsic in constituting the identities of objects is evident at all levels of scientific discourse, including that at which theories describe the entities of fundamental physics. As a consequence, non-eliminative OSR cannot yet be regarded as furnishing any conceptual advance on eliminativism as a proposal for structuralism in the philosophy of science.

Acknowledgments This paper was written for the Banff Workshop on "Structure, Objects, and Causality", and developed significantly at the Sydney Centre for the Foundations of Science. I am indebted to colleagues in both places for inspiration, as well as to audiences for different aspects of this and related work at the Universities of Miami, Canterbury, Auckland, Sydney, Concordia University, the Australian National University, and the annual Canadian Philosophical Association and Dubrovnik Philosophy of Science meetings. Special thanks are due to Philip Catton, David Chalmers, Michael Esfeld, Steven French, Jason Grossman, James Ladyman, Elaine Landry, Holger Lyre, Kerry McKenzie, and Dean Rickles.

References

1. Armstrong, D.M. (1997) *A World of States of Affairs*. Cambridge: Cambridge University Press.
2. ———— (1999) The Causal Theory of Properties: Properties According to Shoemaker, Ellis, and Others. *Philosophical Topics* **26**: 25–37.
3. Bird, A. (2007) *Nature's Metaphysics: Laws and Properties*. Oxford: Oxford University Press.
4. Busch, J. (2003) What Structures Could Not Be. *International Studies in the Philosophy of Science* **17**: 211–225.
5. Cao, T.Y (2006) Structural Realism and Quantum Gravity. In D. Rickles, S. French, and J. Saatsi (eds.), *The Structural Foundations of Quantum Gravity* (pp. 40–52). Oxford: Clarendon.

6. Chakravartty, A. (2007) *A Metaphysics for Scientific Realism: Knowing the Unobservable*. Cambridge: Cambridge University Press.
7. Dipert, R.R. (1997) The Mathematical Structure of the World: The World as Graph. *Journal of Philosophy* **94**: 329–358.
8. Dorato, M. (2000) Substantivalism, Relationism, and Structural Spacetime Realism. *Foundations of Physics* **30**: 1605–1628.
9. Esfeld, M. (2003) Do Relations Require Underlying Intrinsic Properties? A Physical Argument for a Metaphysics of Relations. *Metaphysica* **4**: 5–25.
10. ———— (2004) Quantum Entanglement and a Metaphysics of Relations. *Studies in History and Philosophy of Modern Physics* **35**: 601–617.
11. ———— and V. Lam (2008) Moderate Structural Realism About Space-Time. *Synthese* **160**: 27–46.
12. Floridi, L. (2008) A Defence of Informational Structural Realism. *Synthese* **161**: 219–253.
13. French, S. (1998) On the Withering Away of Physical Objects. In E. Castellani (ed.), *Interpreting Bodies: Classical and Quantum Objects in Modern Physics* (pp. 93–113). Princeton: Princeton University Press.
14. ———— (2003) Scribbling on the Blank Sheet: Eddington's Structuralist Conception of Objects. *Studies in History and Philosophy of Modern Physics* **34**: 227–259.
15. ———— (2006) Structure as a Weapon of the Realist. *Proceedings of the Aristotelian Society* **106**: 167–185.
16. Hawthorne, J. (2001) Causal Structuralism. *Philosophical Perspectives* **15**: 361–378.
17. Ladyman, J. (1998) What is Structural Realism? *Studies in History and Philosophy of Science* **29**: 409–424.
18. ———— (2007) Structural Realism. *Stanford Encyclopaedia of Philosophy*. : http://plato.stanford.edu/entries/structural-realism/.
19. ———— (2007) On the Identity and Diversity of Objects in a Structure. *Proceedings of the Aristotelian Society Supplementary Volume* **81**: 23–43.
20. ———— and D. Ross (2007) *Everything Must Go: Metaphysics Naturalized*. Oxford: Oxford University Press.
21. McKitrick, J. (2003) A Case For Extrinsic Dispositions. *Australasian Journal of Philosophy* **81**: 155–174.
22. Pooley, O. (2006) Points, Particles, and Structural Realism. In D. Rickles, S. French, and J. Saatsi (eds.), *The Structural Foundations of Quantum Gravity* (pp. 83–120). Oxford: Clarendon.
23. Psillos, S. (2001) Is Structural Realism Possible? *Philosophy of Science* **68**: S13–S24.
24. ———— (2006) The Structure, the Whole Structure and Nothing But the Structure? *Philosophy of Science* **73**: 560–570.
25. Rickles, D. (2006) Time and Structure in Canonical Gravity. In D. Rickles, S. French, and J. Saatsi (eds.), *The Structural Foundations of Quantum Gravity* (pp. 152–195). Oxford: Clarendon.
26. Saunders, S. (2003) Indiscernibles, General Covariance, and Other Symmetries: The Case for Non-Reductive Relationalism. In A. Ashtekar, D. Howard, J. Renn, S. Sarkar, and A. Shimony (eds.), *Revisiting the Foundations of Relativistic Physics: Festschrift in Honour of John Stachel* (pp. 151–173). Dordrecht: Kluwer.
27. Shoemaker, S. (1980) Causality and Properties. In P. van Inwagen (ed.), *Time and Cause: Essays Presented to Richard Taylor* (pp. 109–135). Dordrecht: D. Reidel.
28. Slowik, E. (2005) Spacetime, Ontology, and Structural Realism. *International Studies in the Philosophy of Science* **19**: 147–166.
29. Stachel, J. (2006) Structure, Individuality, and Quantum Gravity. In D. Rickles, S. French, and J. Saatsi (eds.), *The Structural Foundations of Quantum Gravity* (pp. 53–82). Oxford: Clarendon.
30. Swoyer, C. (1982) The Nature of Natural Laws. *Australasian Journal of Philosophy* **60**: 203–223.
31. Worrall, J. (1989) Structural Realism: The Best of Both Worlds? *Dialectica* **43**: 99–124.

Index

E.M. Landry, D.P. Rickles (eds.), *Structural Realism*, The Western Ontario Series in Philosophy of Science 77, DOI 10.1007/978-94-007-2579-9,

MIX
Papier aus verantwortungsvollen Quellen
Paper from responsible sources
FSC® C105338

If you have any concerns about our products,
you can contact us on
ProductSafety@springernature.com

In case Publisher is established outside the EU,
the EU authorized representative is:
Springer Nature Customer Service Center GmbH
Europaplatz 3, 69115 Heidelberg, Germany

Printed by Libri Plureos GmbH
in Hamburg, Germany